Road Scars

Place, Memory, Affect

Series Editors: Neil Campbell, Professor of American Studies at the University of Derby and Christine Berberich, School of Social, Historical and Literary Studies at the University of Portsmouth

The Place, Memory, Affect series seeks to extend and deepen debates around the intersections of place, memory, and affect in innovative and challenging ways. The series will forge an agenda for new approaches to the edgy relations of people and place within the transnational global cultures of the twenty-first century and beyond.

Walking Inside Out edited by Tina Richardson

The Last Isle: Contemporary Taiwan Film, Culture, and Trauma by Sheng-mei Ma

Divided Subjects, Invisible Borders: Re-Unified Germany after 1989 by Ben Gook

The Mother's Day Protest and Other Fictocritical Essays by Stephen Muecke

Affective Critical Regionality by Neil Campbell

Visual Arts Practice and Affect edited by Ann Schilo

Haunted Landscapes edited by Ruth Heholt and Niamh Downing

In the Ruins of the Cold War Bunker edited by Luke Bennett

The Question of Space: Interrogating the Spatial Turn between Disciplines edited by Marijn Nieuwenhuis and David Crouch

Spatial Anthropology: Excursions in Liminal Space by Les Roberts

Road Scars: Place, Automobility, and Road Trauma by Robert Matej Bednar

Nature, Place and Affect: The Poetic Affinities of Edward Thomas and Robert Frost 1912–1917 by Anna Stenning (forthcoming)

Road Scars

Place, Automobility, and Road Trauma

Robert Matej Bednar

ROWMAN & LITTLEFIELD
Lanham • Boulder • New York • London

Published by Rowman & Littlefield
An imprint of The Rowman & Littlefield Publishing Group, Inc.
4501 Forbes Boulevard, Suite 200, Lanham, Maryland 20706
www.rowman.com
6 Tinworth Street, London SE11 5AL, United Kingdom

British Library Cataloguing in Publication Information Available

Library of Congress Cataloging-in-Publication Data
Names: Bednar, Robert Matej, 1966– author. | Rowman and Littlefield, Inc.
Title: Road scars : place, automobility, and road trauma / Robert Matej Bednar.
Other titles: Place, memory, affect.
Description: Lanham : Rowman & Littlefield, 2020. | Series: Place, memory, affect |
    Includes bibliographical references and index.
Identifiers: LCCN 2020009552 (print) | LCCN 2020009553 (ebook) |
    ISBN 9781786614131 (Hardcover) | ISBN 9781786614148 (ePub)
    ISBN 9781538148334 (pbk)
Subjects: LCSH: Roadside memorials—United States.
Classification: LCC CC350.U6 B38 2020 (print) | LCC CC350.U6 (ebook) |
    DDC 363.12/5—dc23
LC record available at https://lccn.loc.gov/2020009552
LC ebook record available at https://lccn.loc.gov/2020009553

For Danielle,
who felt this book with me, year after melancholy year,
and for Anika and Lukas,
who have grown up with me always working on this book,
at home and on the road.

Contents

Preface and Acknowledgments

To say that I have struggled to write this book is an understatement. It has challenged me as a scholar and as a person more than I could have ever imagined. It has been as frustrating as it has been fascinating and exhilarating. It has tested the patience of those closest to me, and it has tested my own patience. And it would have never happened without the support of key people at key moments along the way that I want to acknowledge here.

I know some scholars who propose projects and then carry them out roughly as planned, but I have always been the kind of researcher who only discovers what I am doing while I am doing it, the kind of photographer who shoots first and edits later, and the kind of writer who sees writing as a form of thinking that shows me what I think as much as reflects what I think.

But this project has pushed that orientation to its limits. When I started the project, I saw it as a small part of a larger book on the ways people use public material and communicative resources to express their private identities in public space. The working title of that book was *Making Public Space Personal: Everyday Acts of Place-Making*. I visualized a chapter on roadside shines alongside one based on my earlier work on tourist photography as well as several chapters on the more subtle and habitual ways that people carve out space for themselves by temporarily inhabiting dorms and office spaces, developing personalized walking paths in cityscapes and landscapes, and developing a sense of ownership about parking spaces and seating in classrooms. I may still write that book one day, but working on shrines became a consuming passion on my first fieldwork trip to New Mexico in 2003.

The main reason this project stopped me in my tracks in 2003 and set me on the path that ultimately has ended with the book before you now was that I was not prepared for the affective response I would have to roadside shrines. That is, I came into studying them as a scholar of visual, material, and spatial communication and car culture, and not as a scholar of trauma and affect. Unlike the other things I had studied and then planned to study as a scholar, I felt shrines moving me in a confusing, intriguing, and troublingly personal way. It has taken me years to make sense of that feeling, and even longer to channel it into something productive. Even I lost patience with my process as the few years I originally projected to complete it turned into a decade, and especially as a decade approached two decades.

Looking back now, I can see that 2014 was the turning point for me, the fulcrum between me "working on a book" and me "finishing a book." By then, I had completed all the main fieldwork and had been actively writing and rewriting the book for years. As I struggled to find my way through the project, I revised several separate chapters multiple times and revised my overall approach almost as many times. That spring, two key events shook me so hard that they challenged me to change my approach to my work altogether.

The first event was the murder of my fourteen-year-old nephew, Breck Bednar, in February 2014. He was killed in circumstances very similar to adolescent traumas I had myself lived through, many years before. Witnessing his trauma secondhand brought me right back to my own trauma, providing me an urgent personal stake in understanding the mechanisms of trauma, now nourished by a deep engagement in trauma theory and work on my own trauma in therapy. And witnessing Breck's trauma alongside my extended family taught me how different people handle the same secondary trauma differently even as they share the condition of witnessing it together, which is a central theme of the book. Finally, taking a pilgrimage to the site of Breck's murder with his father (my brother) was a powerful experience in its own right, but it also helped me understand, in an embodied way, how troubling it is to both approach and leave a place of trauma. Living with Breck's murder has been impossible, but it has infused my writing with a sense of urgency and taught me an important lesson in how secondary trauma works, particularly when a person's ordinary expectations about the future are violently interrupted, as they are also so often on the road.

The second event was my being denied promotion to full professor later that same spring, which hit me at the time like a sucker punch. The reason given for why I was not promoted was that I had been working on the book so long that publishing it had become the institutional measure of my success as a scholar. I already had spent the intervening three months after Breck's murder processing my trauma and learning to translate its energy into my work, and I came out of the experience of rejection determined to channel my initial disappointment and anger about the decision into motivation to complete the project—to answer their questions about whether I would finish it as well as my own.

But the fact remained that I had neither a completed manuscript nor a book contract, and I truly did not understand how to change that. Thus, while these two events were important, the most crucial thing to come out of them is that my spouse and closest collaborator, Danielle, saw me struggling and asked my old friend and colleague Dean Rader to coach me through to the next phase. In that one stroke, Danielle did what I had never been able to do before as a person and as an academic: to not only recognize when I was drowning but also ask for help when I knew I was drowning. And in responding graciously to my veiled and reluctant cry for help, Dean allowed me to come to terms with the ways that the trauma that was driving my interest in this project was also driving it into an endless tunnel of repetitious revision after revision with no end in sight. Together, Danielle and Dean showed me what I was not able to see: that in analyzing other people's trauma, I was acting-out of trauma instead of working-through it.

From there, we jumped into action. Dean read early drafts and told me that one thing I could not tell myself: "This is important work that people need to see. I can see some things to improve, but it's good enough to show people as it is." More important, he helped connect me to editors he had worked with, who got the project out of my hands for the first time, and into the hands of others. I wish I could say that that was the end of the story, but it was not. It took me three more years of submissions, reviews, and negotiations before I was able to finalize the contract here at Rowman & Littlefield.

Along the way, there have been many other people who have supported me and my work—too many to mention personally, but some whom I must acknowledge directly. First, I have benefited greatly from engaging with audiences where I have presented my work on roadside shrines at academic conferences and symposia as well as at a number of invited "public intellectual" venues. I have presented my work at the national American Studies Association conference as well as regional conferences such as Pacific Northwest ASA and Rocky Mountain ASA as well as at the Western History Association, the American Anthropological Association, and numerous times at the Southwest/Texas Popular Culture Association. Each time, I have learned much from the uncontrolled dynamic of Q&As. Over the years, the most stimulating venues for me were themed conferences and symposia, particularly *TRANSVisual* at the University of Wisconsin in 2006, *Visible Memories* at Syracuse University in 2008, *Contained Memory* at Te Papa Tongarewa Museum in Wellington, New Zealand, in 2010, *Places of Loss* at the Psychoanalytic Institute of Northern California in 2012, *Affective Landscapes* at the University of Derby, United Kingdom, in 2012, *Affect, Images + Digital Media* at the University of Utah in 2015, *SALSA* at the University of Texas in 2018, and *Material Temporalities* at the Max Planck Institute in Göttingen, Germany, in 2020.

The editors I have worked with also have been essential in providing the support and critique I needed to keep the project moving beyond my own limitations. This was true for those who I worked with on edited volumes featuring my early work, such as Dean Rader and Jonathan Silverman, and Anne Demo and Brad Vivian. But I especially want to thank Christine Berberich and Neil Campbell, the series editors for *Place, Memory, Affect* at Rowman & Littlefield, for their support and valuable critical feedback throughout the multiple stages of this project. I first met Christine and Neil when I presented my work at the *Affective Landscapes* conference they organized in Derby in 2012, and since then we have developed a strong working relationship. They worked with me to publish two shorter pieces of mine, "Placing Affect" and "Killing Memory," and have become both enthusiastic advocates and valued critics. The professional editorial staff at Rowman & Littlefield have been excellent as well. Gurdeep Mattu is one of the few academic editors I have worked with who has understood how important balancing words and images is to my work. I especially appreciate the collaborative approach Gurdeep and his staff have taken to both envisioning the book and shepherding it through the pipeline. I would also like to thank the anonymous peer reviewers gathered by Gurdeep, Christine, and Neil. The reviewers' evaluations of a previous draft of the manuscript challenged me to be less abstract and more direct with my argument. I hope they see their input reflected here and agree with me that the process has produced a stronger book.

An even smaller circle of people have had an especially large impact on me and my work. I already mentioned Dean Rader, without whose interventions this project might have languished. My therapist, James Ochoa, compassionately and competently helped guide me as I dug into the deepest anxieties that accompanied my past and present traumas. Renee Bergland at Simmons not only organized the panel for which I wrote my first paper on roadside shrines but also was the first colleague to encourage me to develop it from an article into something bigger. Maria McVarish and Julie Levitt, who I met at the *Visible Memories* conference, brought me to San Francisco in 2012 to present my work at both California College of the Arts and the Psychoanalytic Institute of Northern California. They helped me see the reach of my work and provided me with the opportunity to work directly with both MFA students and psychotherapists. Several colleagues at nearby University of Texas at Austin have been particularly important for me over the years. Josh Gunn from CommStudies brought me in to engage directly with the graduate students in his "Hauntings" seminar in 2009. Katy Stewart from Anthropology not only provided me with the initial impetus to theorize the ways that loose collectives are connected by shared affect in everyday life with her book *Ordinary Affects* but also personally connected me to a stimulating collective of ethnographers and critical theorists at the University of Texas when she invited me to present my work as part of the "Writing and Inscription" Lecture Series in 2012. While attending another event at that lecture series, I reconnected with my old colleague from American Studies at UT, Randy Lewis. Our regular breakfast taco dates ever since have helped create a space for me to be vulnerable with a colleague in a way I only had been before with Dean Rader.

Throughout it all, my colleagues and students at my home institution, Southwestern University, have been the earliest, most visible, and most sustained supporters and critics of this project. I developed my ideas in engagement with my students every day, especially in my Communication & Memory and Visual/Material Communication classes. My colleagues' response to my work in multiple lectures and workshops over the years kept pushing my ideas further. Walt Herbert provided essential advice early on that I only learned how to heed much later. My work to establish and co-coordinate the Situating Place Paideia Cluster brought me into close interdisciplinary collaborations with my colleagues across the university who are engaged in understanding place, especially Laura Senio Blair and Josh Long. Alisa Gaunder, Dean of Faculty, has been particularly supportive as I have rounded the bend toward completing the project. And, of course, my colleagues in my home department, Communication Studies, especially Val Renegar, Lamiyah Bahrainwala, and David Olson, have provided me with a productive environment to work within day to day—a place where I feel known and valued but also challenged to develop a stronger and stronger collaborative practice. Additionally, all of the major fieldwork was funded through competitive internal faculty development funding, and much of the writing was supported by two writing-intensive sabbaticals, so this project literally would not have been possible without my institution's material support.

Of course, none of this would have been possible at all if it weren't for the work of the hundreds of creative and courageous people who have built the shrines I have photographed and analyzed here. Even now, after witnessing thousands of roadside shrines, I still find myself surprised and amazed by each shrine I encounter. I have endeavored to meet each shrine on its own terms while also trying to translate their energy into something bigger. Each time I have witnessed a shrine, I have tried to attune to the feelings it conjures, but I always know I do not feel them the same way as the people who produce them. I hope that my focus on encountering them as a stranger and focusing on their visual, material, spatial, and performative dimensions helps readers appreciate the work that shrine builders have done, even as it brackets off the absent stories of trauma the builders know all too well. I also hope that my efforts to place their work alongside the work of other shrine builders strike the difficult dialectical balance I have sought to enact here, between respecting the radical individuality of each shrine (and each person commemorated and performed through each shrine) while collecting them here to help us all sense the larger cultural trauma they embody.

Finally, my family has been important to the process of producing this book from the beginning. My parents, Gene and Julia Bednar, have been staunch supporters all along, and my extended family in the United States and the United Kingdom has provided me with the foundation I needed to push forward. My wife and children have been most central, which is why I have dedicated this book to them. This book is not necessarily about them, but it is inseparable from my life with them. All of the research and writing I have done for this book has been done with my family close at hand. Danielle and I committed very early on to seek a work/life balance that allowed us to be active and present "attachment" parents to Anika and Lukas even as we have both lived rich professional lives. In the field, they not only were my traveling companions but also my field research team. They helped me spot shrines as we drove throughout the American Southwest and helped me talk through what I was experiencing while on the road. They all still instinctually point out shrines wherever we travel. At home, they have been not only the inspiration to keep moving the project along, year after melancholy year, but also the backdrop for it. Most of the early writing was accomplished either at the shared desk in our living room or in the car parked somewhere during kid naps. My kids have grown up always knowing me as someone working on this project, either on the road doing fieldwork or watching me stationed at the living room desk with my earbuds in, tapping away. By now, I have been working on this project so long that I have just as much of a hard time remembering a time when I was not working on it as they and I do imagining a future when I am not working on it. But here I am, sending it out into the world, welcoming our next adventure.

Chapter 1

Introduction: What Are Car Crash Shrines Doing on the Roadside?

In the summer of 2006, at the end of a long day of driving around the Mojave Desert northeast of Los Angeles, California, I spotted a collection of objects woven into a chain-link fence on the side of a desolate road. From a distance I could make out an array of objects: flattened balloons, photographs, laminated messages, flowers, beads, a Christmas stocking, and a hockey stick (see figure 1.1). I instantly recognized it as a roadside car crash shrine.

Parking on the shoulder and getting out of my car, I went to work doing what I now have been doing across the American Southwest for the past seventeen years: photographing roadside shrines, thinking about how they work as forms of visual and material culture, and working on the book now before you.

As I started to move around the shrine looking for different angles to photograph, I found myself continually drawn to a photograph hanging at the left edge of the assemblage, a photograph of an adolescent boy dressed in a hockey uniform and smiling widely, revealing his braces. The photograph was attached to the chain link fence with string. Just underneath the picture was a hockey stick that looked just like the one in the photograph.

Words were handwritten on the left side of the picture. "Your Dad and I miss having you to enjoy. Glory to God." On the other side of the frame, the same person has written "Love Forever" in quotes and signed it "Mom" with "OXXOXOXX."

About a foot away from this photograph was another photograph. This one was clearly a high school graduation picture. While the hockey picture was laminated, this one was set into a frame, and the frame was surrounded by plastic ivy. It was clear that the picture had been living outside for a while, and that water had gotten in between the frame and the photograph, warping and staining the picture.

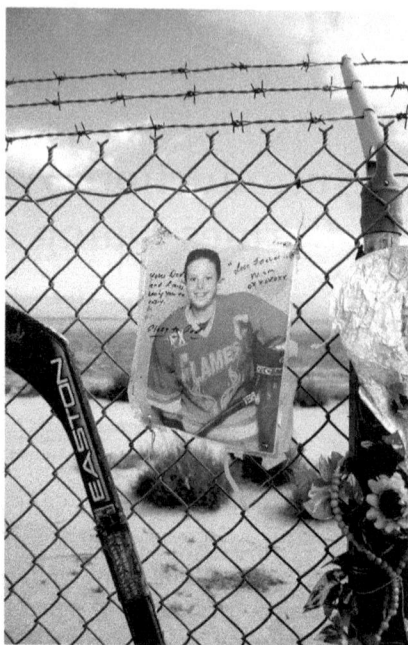

Figure 1.1. East M Avenue @ 50th Street East, Palmdale, CA, USA, July 2006.

Figure 1.2. *Two Portraits*, East M Avenue @ 50th Street East, Palmdale, CA, USA, July 2006.

Studying the hockey portrait and the graduation portrait together, I concluded that the two pictures move forward in time from left to right and represent two moments in the life of a single person. Thinking of the photographs this way, I noticed a significant difference: in the later picture, his braces are gone, and they have succeeding in straightening his teeth. As I stood there ruminating over braces, I was instantly swept into an intense connection with the boy in the picture as well as the "Mom" and "Dad," who are barely out of frame. Suddenly, I was transported to all sorts of places and times in my mind: the portrait studio, the hockey rink, the orthodontist, the graduation ceremony, the ambulance, the cemetery, and so on.

Before long, though, my feeling of imaginatively moving outwards into different times and places reversed and I started to feel all of them converging on the place where I was then standing—connecting the present moment in time, July 24, 2006, to the date of the accident, July 12, 2005, a month after the boy in the pictures turned twenty, and almost a year to the day before my own arrival on the scene.

It was the braces that clinched it for me.

People who do braces are not fatalists. Braces are a cumbersome technology, tangibly embarrassing and physically uncomfortable. To sublimate your feelings of shame and discomfort in the present, you must project a valued benefit in the future. In short, doing braces "takes time" and to do them, you must *believe in the future*. Looking at the hockey and graduation photographs together again, I saw the tragedy materialized in visual form for the first time: The first is a picture of a boy who both believed in the future and believed he was moving toward it. The later picture shows that his projected movement into the future had momentarily come true, as he finished high school and embarked on his adult life. Ultimately, though, the structure of the shrine also shows me that *now*, despite his ordinary expectations and the ordinary expectations of his parents, he is dead, and thus *he has no future* at all.

I imagine that if I saw the picture of the boy in braces anywhere else— within a frame on the wall of someone's home, in a photo album, on the web—I would see it as a representation of the awkwardness of early adolescent masculinity, with its strange brew of pride, shame, hope, and dread. That is, the picture itself is nothing special and has no special appeal to me other than to satisfy a distracted curiosity as someone who studies vernacular visual and material culture. Given the style of the boy's clothes and hair, I would also probably assume that the person pictured in the hockey photograph is still alive, either the same age as he is in the later photograph or at most ten years older, walking around among us now negotiating his twenties.

Seeing it there, though—next to the hockey stick and the graduation picture, situated within a crash shrine, and with the mom's message written across it—it moved me. It first hit me like a tiny trauma: a short, sharp shock.

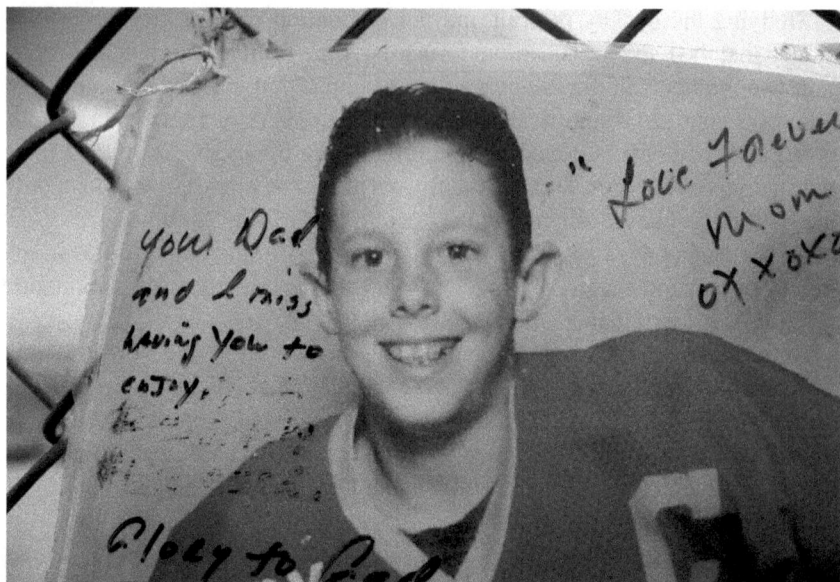

Figure 1.3. East M Avenue @ 50th Street East, Palmdale, CA, USA, July 2006.

But then it quickly turned into a vague but palpable melancholy: a subdued and generalized sadness. This initial shock and the resulting melancholy were not the result of the direct bodily trauma of a car crash fatality, or the trauma of someone who is suddenly aware that they have lost someone they know and love, but a trauma nonetheless.

Traumatic affect emerged most palpably as a recognition of the shrine's complex ontology, demonstrated within the confines of the shrine instead of only my imagination. It took shape most directly in the assemblage of portraits: this person *has been*, this person *was in the process of being* alive when they arrived at this particular spot, but here is where their drive was traumatically interrupted. *Now*, they *are* dead, and this is the particular place where they were transformed from someone who *is* to someone who *was*. This is the place where the person who lived and has been lost is now doubled as a shrine commemorating a person, which itself now lives a life of its own as it *is* on the roadside. There, it is being tended by a mother who is managing not only the loss of her son but also the loss of her ordinary expectations not only of where he would be going in his car that day but also in his life. The shrine not only exists to "keep the young man's memory alive" but also has its own kind of performative agency—an agency I felt acting on me, someone who had never met him.

What moved me that day was feeling the trauma being staged right in front of me, and in front of everyone else driving by. The trauma was located in my embodied encounter with the objects there and the way they were assembled, and in the fact that it was on the roadside. What moved me was realizing that although the shrine first prompted me to imagine the scenes of an absent trauma—the trauma experienced by the young man and the trauma experienced by the mom that precipitated the shrine—I was witnessing a different trauma performed in and through the shrine itself. Because while this and other shrines have all sorts of things done to them—they are built, they are maintained, they are revised, they are contested, they are removed—they also *do something to strangers like me* as they live their lives on the roadside.

The question I have been asking myself since that day is: How can that be? How can a photograph of a stranger stimulate that kind of affect? Even standing there in front of the shrine, I already knew that it was not just triggered by the photograph or even the set of photographs, but the photographs embedded in a larger crash shrine and situated in a radically particular place, so I should rephrase my question differently already: how can a roadside shrine commemorating a stranger make me feel something, much less something that intense? Given the fact that shrines are about death and bereavement, which is always presumed to be heavy with affect, the answer may at first seem obvious, but as a scholar of everyday visual, material, and spatial culture, my job is to interrogate the obvious. It is the *obviousness* of the obvious that gives it its force, so we must work to reflexively extract ourselves from the dynamic long enough to analyze it while still being inside it.

So I keep asking: how can a roadside shrine for a stranger move me? How is it that, when I am encountering a shrine commemorating the lost life of a stranger—a person I have never met and now surely never will meet—I feel strong things in my body that I attribute to the visuality, materiality, and spatiality of shrine instead of only to my own imagination? How can a collection of inanimate objects gathered on the roadside generate affect, especially when they do not *belong* to me or anyone I intimately know? What does a roadside shrine *do* to contain and communicate affect, and how does it do it? And what are these shrines *doing* on the roadside in the first place?

That "doing" is the subject of this book.

What *are* car crash shrines *doing* on the roadside? When I ask this question, I am thinking of it in multiple ways. First, the "doing" that shrines are engaged in is at least twofold: it includes not only the sense of "what are they doing there?" as in, "how did they get there?" and "what purpose do they serve?" and "why are they there?" (especially as opposed to other places they might be, and other cultural forms that might serve the same function of mediating road trauma, especially gravesites or home-based shrines), but also the sense

of "what are they doing there?" as in, "what *work* are they *doing* while they are
on the roadside, for whom, and how?" Further, this second sense of "doing"
splits even farther, into the smaller-scale grief work done at particular sites by
and for intimates (which is well documented and studied by many scholars)
and the larger-scale *cultural* work done by shrines when they act as and are
treated as collective trauma and memory forms by and for strangers (which is
not well studied at all).

Prevalent for decades in Latin America and in the American Southwest,
roadside car crash shrines are now present throughout the United States and
around the world. Although there are many antecedents in many different
cultures that have had similar practices of both burying travelers who die in
transit and marking locations of violent deaths in spaces of mobility, folklor-
ists have established that the specific practice of marking automotive death
sites can be traced back hundreds of years in the Indio-Hispanic American
Southwest. There, the practice is rooted in the traditional practice of marking
the resting places of funeral processions winding their way on foot through
the Sangre de Cristo Mountains, and car crash shrines still are called *descansos* (or "resting places") throughout the Southwest. In the early twentieth
century, not long after the widespread integration of the automobile, the
practice was adapted to memorialize car crash deaths to mark the site of what
Rudolfo Anaya calls an "interrupted journey."[1] Some are simply small white
crosses, almost silent markers of deathsites; others are elaborate collections
of objects, texts, and materials from all over the map culturally and physically, all significantly brought together not in the home or in a cemetery or
mausoleum but on the roadside, in drivable public space, a space of what
Raymond Williams has called "mobile privatization": a public space where
private individuals perform private identities, together, as strangers connected
to each other in a loose kind of collectivity.[2]

In addition to scholarship on spontaneous shrines in general that has studied roadside shrines comparatively (and which we will discuss in chapter 2),
scholarship specifically on roadside crash shrines is robust. It started with a
few scattered articles in the 1990s followed by a widening and deepening
literature starting in the early 2000s that continues to grow into the present.
Early work either focused on case study approaches to localized instances of
the practice or analyzed it in very general terms while drawing connections
to previous analogous cultural forms.[3] In the early 2000s, the field grew
and diversified quickly, generating a number of foundational articles and a
monograph by Holly Everett, *Roadside Crosses in Contemporary Memorial
Culture* (2002), which established roadside shrines as an object of study and
still serves as a comprehensive introduction into the practice, particularly in
the United States.[4] This early scholarship established the general point that
shrines are vernacular forms built by ordinary people to memorialize ordinary

people lost in car crashes, and do so in public landscapes where strangers encounter them.[5] By the 2010s, scholars in a diverse set of disciplines and interdisciplines, including me, began publishing work that looked more closely at how these dynamics work in detail.[6] Along the way, scholars have analyzed the phenomenon in a number of cultures, showing important cultural differences in how the practice is evolving around the world, especially in Australia, the Czech Republic, Ireland, the Netherlands, New Zealand, Poland, Romania, Sweden, Russia, and the United Kingdom, including how the practice is being mapped to more indigenous local bereavement and place-making practices while still being a global phenomenon.[7]

Scholars working on roadside crash shrines in a number of different disciplines have focused mostly on the producers and direct users of the shrines: the intimates who already knew the person killed in the crash, the people who can remember who was lost, and who work with crash shrines because they need an embodied way of negotiating their traumatic grief. As these scholars have established, the primary function of such shrines is to create a performative space for *mourning* with a secondary function of providing a potential *warning* to other drivers who encounter them. Even then, most of this scholarship has focused on the mourning function and on documenting mourning and memorialization practices within microcultural contexts while making broad assertions about the way that these shrines "bring private grief into public space." Little of the published work has effectively theorized or analyzed the larger-scale importance of roadside shrines within contemporary discourses of memory, trauma, and (especially) automobility. There is even less work on the processes by which both mourning and warning functions work for both strangers and intimates at particular sites as well as how these functions are embedded within the larger dynamics between the individual and collective experience of trauma—especially road trauma, which has distinctive characteristics that distinguish it from other forms of trauma.

The work of Jennifer Clark and her collaborators stands out in this regard, representing a sustained body of work that engages many of the larger-scale questions that I engage in this book, but mostly as they manifest in Australia and New Zealand.[8] In "Authority from Grief," for instance, Clark and Majella Franzmann argue that shrines are not simply about gaining authority through memory and mourning but about gaining authority by seeking *public acknowledgment* of personal grief. Clark and Franzmann argue that a shrine accomplishes a "transformation of the roadside into sacred space . . . because memorial makers find authority in their experience of grief, presence and place."[9] They also suggest that the practice may "be part of a fight against the depersonalization process of modernization and urbanization."[10] In an earlier work, Clark and Franzmann located this "fight" squarely within automobility, as I am doing here as well. They argue that roadside shrines represent "the

only way to register and put into public debate repeated road death as the dis-
turbing outcome of automobility"; and even then, shrines exist as "an accu-
mulation of small crashes" that have "the numbers" but not "the purpose" to
elevate them or collect them into a palpable group.[11] Finally, in a later single-
author work, Clark argues that shrines "can challenge us to broaden our idea
of motoring heritage" to include the "dark side" of the cultural history of the
automobile and car culture into a collective heritage, or cultural memory, of
both the benefits and costs of automobile-centered mobility.[12]

Scholars continue to assert that roadside shrines address "the public," but
have not fully analyzed or theorized how these shrines actually *produce* the
more differentiated publics they address. An exception here is Elaine Campbell.[13] Although Campbell uses roadside memorials more as a case study to
help explain how new forms of public spheres are produced rather than as
an object for her to analyze directly, her perspective on roadside memorials'
capacity for forming temporary and tenuous kinds of collectives is similar
to the main argument I make in this book. Campbell argues that roadside
memorials are exemplars of contemporary emergent public spheres because
they function "as a connective and generative space which reflexively nur-
tures and assembles the cultural politics (and political cultures) of which it is
an integral part."[14]

My work here explores these important dimensions of roadside shrines
as well, not only characterizing them in fairly general terms as Clark and
Franzmann and Campbell have done but also showing in extensive detail
how they operate within the particular cultural technologies and practices
present at specific crash shrines to have a certain collective agency in the
culture. This is different from the agency they give to the intimates who work
though their grief at the sites, which is an important distinction. For instance,
Catherine Collins and Alexandra Opie characterize roadside shrines as places
with agency, but they see that agency as a function of the grief work done
by intimates at shrines and not any larger-scale force or agency the shrines
themselves may have as culturally charged objects in public space. Collins
and Opie argue that such shrines provide a space for working-through the
violent deaths that occur there by creating an orderly inversion of "the chaos
of traumatic memory and grief," giving the people who build and maintain
shrines a sense of agency in the face of traumatic loss.[15] This is also a central
claim of Jennifer Clark and Majella Franzmann, who see roadside shrines as
a way for mourners to claim an "authority from grief" in the public sphere.[16]

Claiming the identity of a grieving subject in public is certainly a primary
function of any roadside shrine. Whatever else is communicated by the
shrines, one thing rings through clearly: that the people who build shrines
in the public right-of-way are doing so for a reason and think that they have
the right to do so. They think that there *should be* room for them not only

to practice their acts of private remembrance in public space but also to be recognized as legitimate in doing so. But, as we will see, as shrines do this, they also do much more.

Most of all, what is not established is an understanding of the processes by which the mourning and warning functions work as collectivizing forces for both intimates *and* strangers, separately and together. The purpose of this book is to address this gap by analyzing the visual, material, and spatial dimensions of roadside shrines that draw together strangers and intimates into a loose kind of collective "public" that has the potential to know road trauma as a *cultural* trauma, not simply a collection of small private individual traumas performed in public. Put simply, what crash shrines do is inscribe traumatic affect into the landscapes of automobility, challenging strangers to remember that the everyday, ordinary traumas of automobility continue to haunt the automotive landscape as part of a much larger disavowed cultural trauma.

Indeed, in a culture more and more focused on individual and collective trauma and affect, roadside shrines are iconic places of public feeling— landscapes where affect collects and is encountered. However, to say that affect is collected and encountered in a landscape is not the same as defining which varieties of affect are collected, encountered, and performed there, and how they are communicated to witnesses, if at all. Roadside shrines are like touchstones—powerfully affective to many, but in unpredictable, undetermined, and uncontained ways, and in different ways to different people. Affect itself is always radically uncontained and only obliquely related to the cultural frames of reference we use to make sense of it. Patricia Clough argues that affectivity is "a substrate of potential bodily responses, often autonomic responses, in excess of consciousness."[17] This is particularly the case in the experience of affect associated with death, that ultimate apotheosis and negation of affect, and even more the case with the way death is placed.

I first began to apprehend the complexity of these affective dynamics at the shrine featuring the picture of the boy and his braces next to his weathered graduation picture. I had been visiting and photographing shrines long enough by then to know that I was feeling *something* in my encounters, but this was the first time I recognized that I didn't really understand exactly *how* and *why* I was feeling not only the "something" I recognized but also "something else"—something more and different than what I was able to analyze directly up to that point as a scholar of landscape and visual culture. As we will see, that something else is the affect produced performatively through visual objects and spaces emplaced on the roadside over time.

That recognition has driven my work ever since, and has led me to the analysis and conclusions I will share throughout the book. For me, one of the most fascinating things about roadside shrines is the way they traffic in both the absolutely material and the absolutely intangible at the same time. My methodological goal in analyzing them, therefore, has been to theorize while staying close to the ground. I explore questions that always must be rooted to empirical realities on the ground—the documentary representation and analysis of actual shrines located in actual places on the roadside that I have interacted with and photographed in certain embodied field encounters—but whose meanings, practices, and structures are located "elsewhere," in discursive phenomena that cannot necessarily be pictured in photographs but that are nonetheless present. This has led me to a wide and deep engagement with theories of visual, material, and spatial communication, and it has led me to a wide and deep engagement with theory and research in the fields of trauma studies, memory studies, and automobility studies.

I will critically engage these scholarly conversations more extensively throughout the book, but I want to provide a quick orientation here. My work on shrines is located at the convergence of many interdisciplinary "turns" in contemporary humanities and social sciences scholarship: visual, spatial, material, affective, and nonhuman. In keeping with my engagement with affect theory, I prefer the term "attunement" to "turn" because a turn implies a turn *toward*, which also implies a turn *away from*, but one can attune to multiple things at once.[18] These attunements have led scholars like me to critically engage nonhuman, more-than-human, nonrepresentational, and extralinguistic cultural forms. I take what Victor Buchli calls a multisensory and performative approach to "the phenomenological and somatic effects" of visual, material, and spatial culture "beyond textuality."[19] Moreover, my work grows out of my long interdisciplinary engagement with visual studies, material culture studies, and cultural geography, so my approach to road shrines would be best described as visual/material/spatial. My approach emphasizes the interconnectedness of the communicative dimensions of the visual, the material, and the spatial as they are embedded in everyday cultural practices, which themselves are always embedded in larger discursive formations. I am always more concerned with what images, material objects, and places *do* and what people *do with* them than I am with what they *represent* and *mean* as texts or *are* as pictures, objects, or places. I am most concerned with how images and objects surface in the world at certain places within and between bodies and objects, and within certain physical and cultural fields—and how we live with them.

Similarly, cultural geographers Gillian Rose and Divya Tolia-Kelly argue for an approach to scholarly analysis of images, objects, and spaces that analyzes and theorizes the interconnectedness of visuality/materiality to "account for the embodied politics present in the everyday material world," where "the 'visual' and the 'material' should be understood as in continual dialogue and co-constitution."[20] Such an approach acknowledges the representational function of visual/material culture, but emphasizes the social practices and performative and intersubjective relational fields within which visual objects are performed, maintained, and contested. That means thinking about visual objects as things that do something and have things done to them, not simply as texts working through structured signification. Here, in the realm of both affect and discourse, things communicate without "speaking," and witnesses can sense and feel and understand without being "told." All the while, the gaps and slippages in intersubjective communication between people and pictures, objects, and spaces persist, because visual, material, and spatial communication is never entirely representational, and thus never contained in the realm of interpretive "meaning," but embodied in the physical, performative encounters between and among bodies and matter.

Likewise, I have turned toward affect theory because it helps me think through what Lawrence Grossberg identifies as "a gap between what can be rendered meaningful or knowable and what is nevertheless livable" or sensible: what I will be calling that elusive "something else" that seems to be present at roadside shrines.[21] Engaging affect in the context of memory and trauma will make sure that we apprehend the always partial and problematic dynamics of experiencing and communicating trauma. My thinking about place and landscape is similarly focused on processes, gaps, and material practices. In *Landscape and Power* (1994), W. J. T. Mitchell argues that understanding landscape means thinking of landscape as "a dynamic medium"—"not as an object to be seen or a text to be read, but as a process by which social and subjective identities are formed" at certain unique but interconnected locations.[22] In short, analyzing landscape as a medium asks "not just what landscape 'is' or 'means' but what it does, how it works as a cultural practice."[23] And, finally, as a cultural practice, roadside shrines are also thoroughly embedded in what John Urry calls the "system of automobility."[24] More broadly, this "system" of automobility is better understood as what Michel Foucault would call a "discursive formation" or "apparatus." Foucault argues that discursive formations do not have material form themselves; instead, they are manifest in the "practices that systematically form the objects of which they speak," where "speaking" includes a number of enunciations other than literal speech, including visual, material, and spatial practices.[25] Engaging Foucault here will help us understand that automobility is more than a system of transportation; it is a diffused but productive cultural logic that organizes subjectivities, bodies, objects, and practices toward autonomous individual mobility as an explicit and implicit cultural value.

While the book is deeply and widely engaged with critical theory, the book is grounded in empirical field research, which is particularly evident in the thousands of photographs I have taken of hundreds of crash shrines in the seventeen years I have been doing my mobile fieldwork. That fieldwork has oscillated between continuing documentation of numerous Central Texas sites near my home in Austin, Texas, and a number of more extensive research trips to sites scattered around the American Southwest: trips to New Mexico in 2003, 2006, 2010, 2012, 2016, and 2019, to the Texas side of the Texas/Mexico border in 2004–2005, and my most extensive work throughout Texas, New Mexico, Arizona, Colorado, Utah, Nevada, and California in 2006.

When I first started planning this book, I was most curious about what shrines meant—what they represented. The first few shrines I visited contained so many different objects that were grouped together in such odd ways that I initially wanted to "crack the code" to figure out why people built them the way they did. Consequently, I imagined I would use a comprehensive ethnographic process of documentation, where I would not only photograph different shrine sites and make field notes about my encounters with the shrines but also do research on the victims to learn how they died and then do interviews with the people who built and/or maintain the shrines.

The more existing scholarship I read on roadside memorialization that took a similar approach and the more shrines I visited in person, however, the more I was drawn to developing a different approach, one that both extended the scope of existing research on shrines and better drew on my own strengths as both a photographer and an interdisciplinary scholar of material and visual culture in shared landscapes. It's something I had learned long ago as a photographer: whenever I find that I am replicating someone else's picture, I either reposition the camera in relation to the object or change the object I am photographing. In this case, looking at the excellent work that was being done on how and why people build shrines and how shrines served as the location of ongoing griefwork that helped them work through the loss figured in the shrines, I repositioned the camera to focus on the physicality of the shrines themselves. And the longer I did this, the more I was also able to focus on how that physicality changed over time, which has added a longitudinal temporal dimension to my work as well, inadvertently aligning it with emerging models of "slow scholarship."[26]

The more attention I paid to the visual, material, spatial, and temporal elements themselves, and the more I attention I paid to the actual roadside locations of the shrines, the more fascinated I became with how they worked as visual/material/spatial objects for strangers driving by, independently of how their producers used them. From what I have seen in manuscript reviewer comments over the years and Q&A sessions wherever I present my work, I know it can seem puzzling to some people, but once I decided that I was most interested in how shrines directly communicate to strangers as visual objects and places, I gave myself the discipline of deliberately encountering shrines *as a stranger myself* wherever possible. That meant that I would encounter them deliberately not knowing the stories of the actual crashes and information about the victim's identities, and deliberately not knowing why and how the shrine producers had selected and arranged objects in their shrines. I found that when I knew the story of the crash behind a shrine, I would read it back into the shrine through projective imagination and mistake it for coming from the shrine. To better understand how shrines can communicate to strangers, I wanted to know what I would see, think, and feel if I encountered only things I could see, touch, and move myself around, so I gave myself the discipline of encountering them entirely as visual, material, and spatial forms. In short, the trauma I have focused on is the trauma present at and performed through the shrines themselves, not the spectacular but absent trauma that made the shrines necessary.

Finally, to better understand how shrines speak to and from the culture of automobility, I developed an *auto-mobile fieldwork* methodology that ensured that I kept focused on the way almost everyone who encounters a roadside shrine in person does so: from the literal and metaphorical frame of a car windshield. Where traditional ethnographic fieldwork often presumes a singular and static location, the field I have been studying is a space of mobility— a distinctive physical and social location where people move through but do not dwell. Accordingly, when I am doing mobile fieldwork, I am not traveling to or dwelling within a set of predetermined locations. Sometimes, I am looking for a particular shrine that a friend or colleague has told me about or that I have learned about in media coverage. Other times, I am following my own field notes or memory back to sites I have encountered before. Most of the time, though, I am simply driving around, eyes peeled, on the lookout for anything resembling a cross on the roadside. That's how I first encountered the site with the boy in braces, and how I first encountered most of the sites you will see pictured in this book. I intentionally designed this method to approximate the phenomenological way other drivers "accidentally" encounter shrines while driving. This was especially true once I began to see this accidental character as an analogue to the way trauma is repeatedly experienced as an intrusive experience. The main difference, of course, is that I am actually seeking out shrines to "accidentally" encounter. And the other difference is that instead of driving by, I get out of my car to meet the shrines on the ground. But even then, I do so just steps from my car, and thus just steps from my subjectivity as a driver.

As you may imagine, driving around in a car looking for shrines devoted to people who die in car crashes is a peculiar practice. Some shrines scream out to be noticed even at highway speeds. Others blend into their surroundings and are easy to overlook among other more institutional road signs or are obscured by trees or brush. I have seen thousands of shrines since I started this project in 2003, but I know from recursive traveling that I must have missed a lot as well (and sometimes see ones once that I cannot find later).

I am a little jumpy about it: I often mistake fire hydrants, gas line markers, political campaign signs, real estate signs, and fiber optic cable markers for crosses, for instance. I have stopped to photograph hundreds of shrines over the years, giving me thousands of photographs and hundreds of pages of field notes to think through. And even then, I have many other notebooks filled with location notes for many other shrines that for a number of reasons I decided to pass by. I call them my "DNS" notebooks: "Did Not Stop."

There is a stretch of State Highways 85 and 86 in the Tohono O'odham Indian Reservation in southern Arizona where there are over one hundred shrines in less than one hundred miles of road.[27] In other places, I will drive for hours without seeing a shrine. Driving this way, part of me hopes that I will see no sites at all, that no one else has died, that everything is going to be OK for just a moment or a mile longer. But of course part of me also hopes there are more—for "research purposes"—because that is why I am driving on that road in the first place. And, unfortunately, on the roads of the American Southwest, there are lots of opportunities for me to feel this conflict in my gut.

Compared to the commercial communication that surrounds them and the other more official and elaborate forms of roadside cultural memory near them—such as statuary, historical markers, and named buildings, streets, etc.—a roadside shrine is particularly mute. But it does carry a certain power: the power of "spatially anchored" material self-evidence of trauma.[28] Indeed, as we will see later in the book, the material situatedness of a crash shrine—its location at or near the location where someone died on the road—is its primary claim for authority as well as its function as a portal between the living and the dead. This material self-evidence also forms the foundation of the self-evident material appeal to undeniability it makes upon passers-by as well: *here, right where the shrine is, something terrible happened, and this shrine simultaneously represents and performs that fact.* And this is how roadside shrines materialize another memory/space as well: a space for secondary witnessing of the everyday traumas embedded within a culture that lives in and through cars.

Created as transference objects for the bereaved, crash shrines live and die also as transference objects for anonymous drivers as well, who encounter shrines without knowing the people memorialized, without being "inside" the micropublic who maintains a social presence for the victim by commemorating a specific life lost, but being contained inside a different public: a motoring public made aware of lost fellow drivers, and made aware that this one road trauma is part of a much larger but unrecognized trauma.

Seen this way, the mourning and warning functions of roadside shrines converge: they do important work not only for the individuals who mourn the loss of *their* loved ones but also for the larger collective, not only as a warning but also as an implicit assertion of *affiliation*—an assertion that *my* trauma is *your* trauma, too—or, more importantly, *our* trauma. If the memory of the lost object were not performed for and within a public of strangers, there would be no knowledge of the loss, but a crash shrine always carries with it an implicit ethical demand: *my loss is our loss, and if you are driving on this road when you see this shrine, you are part of that "our."* Roadside shrines can help us remember that we are part of a collective—in this case, a public that knows that disavowed traumatic automotive loss exists. But because they do so in the spaces of automobility, they can also help us remember that the loss is a collective loss for a culture dedicated to living with and through cars: they address a public that will keep on driving past them, into a projected future. Ultimately, as I will demonstrate throughout the book, at the center of it all is an assertion that all these dispersed individual traumas add up to a *cultural* trauma.

ROAD SCARS

Roadside car crash shrines are a kind of scar on the physical and discursive landscapes of automobility: a *road scar*. They give form to a massive cultural trauma we have all endured in our entanglement with cars and car culture now for over a century but have not figured out how to work through. I invoke the figure of the scar because a scar marks the presence of a traumatic wound in the process of healing. It is a healing wound that leaves a material trace, which persists from the past into the present, like a relic. A scar is a physical remainder that serves as a reminder not only that a past trauma happened but also that it continues to be present *as* a trauma, not simply *as a memory of* a trauma.

Clearly, in the early days of their lives as crash shrines, most of these sites look more like an open wound than a scar—quickly appearing, metastasizing rapidly to gather an intense array of objects placed there by a large number of different people, and changing both form and content rapidly. Over time, though, as a shrine settles into the landscape, it looks and acts more like a scar because a scar simultaneously marks a space of trauma *and* shows that some healing has occurred at the site of the original trauma. That's because both the trauma *and* the healing have left a trace on the surface of the body or the landscape. And while a wound calls out for treatment, a scar calls out for witnesses—demanding that a "we" sees and feels its material presence, and demanding that a "we" sees itself as part of the public it addresses even if we ignore it.

Which brings me back to the central cultural function of roadside shrines: to give material form to the cultural trauma of car crash deaths. When

I photograph and analyze a shrine, my questions are not "What does this shrine mean?" or "What do shrines mean?" but instead "What is this shrine doing? How does it do it? To what effects? How is that related to what road-side shrines do in general? And what claims do these shrines make on publics as they do what they do on the roadside?" Simply putting a shrine on the road-side does not create collective recognition or memory of a cultural trauma. For a loss to be realized as a collective loss, it must both *address* a public and be *witnessed* by the collective in some kind of performative encounter that enacts a collective. Where individual trauma is borne individually, and mass trauma is borne socially, massive but dispersed, individual traumas like car crashes are always held in suspension between the individuals who directly perform them and the publics who witness the testimony of traumatized individuals. Indeed, witnessing itself is the practice by which collective aware-ness of individual trauma is performed, and by which collective memory happens. As we will see, showing and witnessing road scars is not only an act but also a performative social practice: something people do together in patterned ways, whether they know they are together and that their acts are patterned or not. Such a performance is an ongoing embodied, intersubjec-tive, performative encounter located in unique spaces and at particular times.

To better demonstrate this fact, I have tried to bring that performative social practice of witnessing more directly to you as you work your way through the book. You will see it most directly in the way my words and pictures interact. Part of that comes from the fact that I am both a scholar and a photographer. My work is part of a growing body of work by scholars of visual communication and visual culture who not only *analyze* visual communication but also *produce* visual media. But it is also because I have designed the book to work as what W. J. T. Mitchell calls an "imagetext."[29] Since the late 1990s, particularly after W. J. T. Mitchell's *Picture Theory* (1994) developed a theoretical framework for understanding different word/ image relationships in print media, there has been a growing interest among scholars of visual culture, material culture, visual anthropology, visual sociol-ogy, visual rhetoric, cultural geography, landscape studies, and architecture/ design studies in producing scholarly works that use words and images equally to communicate their argument.[30] Unlike most scholarly books that privilege the written word and use images to illustrate the argument, and unlike most photography books that privilege the visual and separate it from the written word, my work strategically alternates between synchronizing and syncopating pictures and words to ensure that the pictures and words function as equal but distinct partners in the work of communicating my argument.

For me, photography is a central feature of my methodology: making pictures is not only the *product* of my analytical project or something that *illustrates* my analysis but also a *means* of analysis itself. As you have already seen in this opening chapter, I sometimes use photographs to provide illustrations of the visual phenomena I am analyzing and sometimes directly analyze the photographs I display, but I also use visuals as condensed *visual arguments* that run parallel to the written argument. As such, they function as a more sensorially direct method of analyzing the things being analyzed as well as *visually demonstrating* the analysis. If you took away the pictures, the analysis would not work, but the same is true for the words. The words and images are designed to work together to carry the argument equally in a kind of syncopated show-and-tell dynamic to demonstrate the analysis and observations together in the same textual space.

This move to word/image analysis and argumentation has been made more frequently among scholars like me who do research on affect, who critically use and theorize photographs as a way of restaging the scene of an affective encounter with images, objects, and spaces more directly for the reader. What photographs do is virtually bridge the gaps between the materiality of the space and time of the photographed scene and the subsequent spaces and times of different photographic encounters. My photographs seek to engage something that cannot be pictured in them, but nonetheless is present: the troubled and troubling affects that get staged at and through roadside shrines. Knowing the affordances and constraints of photography as well as I do, I nonetheless bring them to you—not to transparently *show* you something but to *share* them with you to invite you to move toward them with me, from your own space and your own trajectory, for sure, but nonetheless together, simultaneously.

Road Scars is fundamentally a transdisciplinary and transmedia endeavor, working not only at the intersection of visual culture, material culture, spatial studies, trauma studies, memory studies, and automobility studies to build a synergistic approach to analyzing the pictures, objects, and spaces that make up roadside shrines but also working to develop a distinctive approach to performing and communicating research through the connections and gaps between my written text and original photography. In words and pictures, the book shows in detail the lives led by these road scars, where they produce a distinctive set of publics and thus a distinctive kind of public affect and memory: one significantly tuned not only to the frequency of contemporary politics and poetics of traumatic individual and cultural memory but also to the frequency of automobility. A crash shrine is no generalized *momento mori*, one which would seek to remind people that they, too, one day somehow will die. This *momento mori* is *targeted*; it reminds *people in cars* that they, too, could *die in cars* or *die because of cars*. Because roadside shrines by definition only exist within a landscape circumscribed by automobility, each one not only speaks to and from an existing discursive collective—automobility, and the motoring public within it—but also activates and constructs a new collective: a *knowing* motoring public that *knows* that accidents are embedded within automobility.

The key to understanding crash shrines as targeted *momento mori* is seeing them not only as individual acts of remembrance or as localized mediations of trauma but also as ongoing performative acts making some claim to a larger collective as well. Doing that means doing what I have done in this book: analyzing them as visual, material, and spatial phenomena while also situating them within a dynamic of intersecting contemporary discourses—trauma, memory, automobility. There, the dispersed and individual acts of road trauma, memorializing road trauma, *and* experiencing other people's acts of memorializing road trauma all mirror each other as they intersect on the road to provide evidence of and to mediate a massive cultural trauma that is only now becoming sensible a little over a century after the widespread integration of automobiles into our lives.

Showing how these complex dynamics work will require detailed attention to the multiple physical, performative, and discursive elements of roadside shrines, and will require an analysis of how all of them work together to produce the effects they do. Roadside shrines materially assert the continuing presence of missing persons, and such a performance is not only always *materialized*—located in physical objects and the relations among them—but also always *territorialized* and *temporalized*—rooted in a particular kind of landscape for a certain time. There, they are located in *some things* that are *placed together* in a *somewhere* for *some time*, which are always also located within cultural discourses. For roadside shrines, that somewhere is always embedded within the everyday landscapes of automobility. There, bodies, objects, spaces, and cultural discourses intersect at road shrines to perform traumatic affect and memory in four main ways—*place-making, materialization, performance*, and *interpellation*.

After setting up the larger framework for the study in chapter 2, where we will situate roadside shrines within discourses of trauma, memory, and automobility, we then will look at each of these four dynamics in detail in the separate chapters that follow. Together, the processes work to visualize, materialize, and spatialize road trauma to address both intimates and strangers in similar but different ways. How these four processes interact at particular sites is what gives each site its distinct identity.

For now, let me first define these main processes to orient you before we move forward. First, road trauma shrines *make places* in certain unique spatial locations, at (or as close as possible to) the actual place where an individual or group of individuals died in an automotive accident (see figure 1.4). Because these crash shrines are located in the public right-of-way, where they involve carving out space for the shrine within an existing place, the shrines are usually deliberately set off from their surroundings in some way. Thus,

Figure 1.4. *Shrines make place by enclosing an inside from an outside, evidencing the crash, and establishing relations of proximity and distance with the surrounding public landscape.* **New Mexico State Highway 518-South, North of Las Vegas, NM, USA, December 2010.**

part of emplacing a site means enclosing affect and memory into an inside separated from an outside, which sets up complex implicit and explicit relations between the site and its surrounding landscape. Some sites have very precisely established and maintained boundaries, where others flow in all directions from a central piece of the automotive infrastructure or central memory object. This is the subject of chapter 3.

Second, once the site has been established as a site, it accumulates and magnetizes things so that anything included in it *materializes* some site-bound effort to contain and communicate affect to the dead directly, to other survivors, and a larger set of publics through transference and relations of proximity (see figure 1.5). This is the subject of chapter 4.

Third, once a shrine has been enclosed and emplaced, the shrine becomes a platform where shrine builders and visitors *perform* trauma *over time* by using the shrine as a kind of portal to communicate to the dead and by adding, moving, removing, or replacing things within the site, and by revising the overall design of the site as the site is negotiated over time (see figure 1.6). This process leaves its own material traces that are readable even by strangers, which will be the main thing we will analyze here, but it also

Figure 1.5. *Shrines serve as proxies for the dead by materializing affect and memory through both transference and magnetized relations among things.* New Mexico State Highway 518-North, South of Taos, NM, USA, December 2010.

Figure 1.6. *Shrines mediate trauma as they are performed over time through material and spatial practices.* Farm-to-Market Road 1325-North @ Loop 45-East, Round Rock, TX, USA, July 2003 and November 2012.

includes more explicit forms of communicating with the dead and others who encounter the site, particularly written messages. This is the subject of chapter 5.

Fourth, no analysis of road trauma shrines would be complete without an analysis of how these sites *interpellate* a public of strangers within what Lauren Berlant calls the "intimate public sphere" to create a particular kind of loose and transitory public structured in and through what Gillian Rose calls the "collective experience of 'feeling'" that comes from participating in "the motoring public" as a loose form of sociality that witnesses shrines together (see figure 1.7).[31] This is the subject of chapter 6.

Taken together, these first six chapters show how road trauma shrines anchor trauma to places and objects that are magnetized to each other through relations of proximity to produce sites that function as communication platforms, communicating through visual, material, and spatial means to the motoring public in an experiential form. The book ends with a final chapter that concludes the book by analyzing in detail the six-year-long life of a single shrine site to try to account for the "remaining" affective dimensions of road trauma shrines. There most of all, I focus on what it will mean

Figure 1.7. *As shrines are placed, materialized, and performed, they interpellate multiple publics, including a loose collective called "the knowing motoring public."* **U.S. Highway 290-East, Elgin, TX, USA, March 2015.**

for us to take up the burden that shrines are performing—what it will mean to remember the past road trauma of strangers as we continue to drive into the future.

NOTES

1 Rudolfo Anaya, Denise Chavez, and Juan Estevan Arellano, *Descansos: An Interrupted Journey* (Albuquerque, NM: El Norte, 1995). See also Michael Trujillo, *Land of Disenchantment: Latina/o Identities and Transformations in Northern New Mexico* (Albuquerque: University of New Mexico Press, 2009).

2 Raymond Williams, *Towards 2000* (London: Hogarth Press, 1983), 187–89.

3 See Anaya, Chavez, and Arellano, *Descansos*; Alberto Barrera, "Mexican-American Roadside Crosses in Starr County," in *Hecho en Tejas: Texas-Mexican Folk Arts and Crafts*, ed. Joe S. Graham (Denton: University of North Texas Press, 1997), 278–92; Kate V. Hartig and Kevin M. Dunn, "Roadside Memorials: Interpreting New Deathscapes in Newcastle, New South Wales," *Australian Geographical Studies* 36, no. 1 (1998): 5–20; Cynthia Henzel, "Cruces in the Roadside Landscape of Northeastern New Mexico," *Journal of Cultural Geography* 11 (1995): 93–106; Holly Everett, "Roadside Crosses and Memorial Complexes in Texas," *Folklore* 111, no. 1 (2000): 91–118; David Kozak and Camillus Lopez, "The Tohono O'odham Shrine Complex: Memorializing the Location of Violent Death," *New York Folklore* 17, no. 1–2 (1991): 10–20; George Monger, "Modern Wayside Shrines," *Folklore* 108 (1997): 113–14; and Robert James Smith, "Roadside Memorials: Some Australian Examples," *Folklore* 110, no. 1 (1999): 103–5.

4 Holly Everett, *Roadside Crosses in Contemporary Memorial Culture* (Denton: University of North Texas Press, 2002).

5 Jennifer Clark and Ashley Cheshire, "R.I.P.: A Comparative Study of Roadside Memorials in New South Wales, Australia and Texas, USA," *Omega* 35, no. 2 (2003–2004): 229–48; Jennifer Clark and Majella Franzmann, "Authority from Grief: Presence and Place in the Making of Roadside Memorials," *Death Studies* 30, no. 6 (2006): 579–99; Charles Collins and Charles Rhine, "Roadside Memorials," *Omega: The Journal of Death and Dying* 47, no. 3 (2003): 221–44; Jeff Farrell, "Speed Kills," *Critical Criminology* 11 (2002): 185–98, 185; Rebecca Kennerly, "Getting Messy: In the Field and at the Crossroads with Roadside Shrines," *Text/Performance Quarterly* 22 (2002): 229–60; Jon K. Reid and Cynthia L. Reid, "A Cross Marks the Spot: A Study of Roadside Death Memorials in Texas and Oklahoma," *Death Studies* 25 (2001): 341–56; and Anita Torres Smith, "Descansos: Markers to Heaven," *Journal of Big Bend Studies* 12 (2000): 259–70.

6 Robert M. Bednar, "Making Space on the Side of the Road: Towards a Cultural Study of Roadside Car Crash Memorials," in *The World is a Text*, 3rd edition, ed. Jonathan Silverman and Dean Rader (Upper Saddle River, NJ: Pearson, 2009), 497–508; Robert M. Bednar, "Materialising Memory: The Public Lives of Roadside Crash Shrines," *Memory Connection* 1, no. 1 (2011): 18–33; Robert M. Bednar, "Denying Denial: Trauma, Memory, and Automobility at Roadside Car Crash Shrines," in

Rhetoric, Remembrance, and Visual Form, ed. Anne T. Demo and Bradford Vivian (London: Routledge, 2011), 128–45; Robert M. Bednar, "Killing Memory: Roadside Memorials and the Necropolitics of Affect," *Cultural Politics* 9, no. 3 (2013): 337–56; Robert M. Bednar, "Placing Affect: Remembering Ordinary Trauma at Roadside Crash Shrines," in *Affective Landscapes in Literature, Art, and Everyday Life*, ed. Christine Berberich, Neil Campbell, and Robert Hudson (Furnham: Ashgate, 2015), 49–67; Rachael M. Byrd, "Rest in Place: Understanding Traumatic Death Along the Roadsides of the Southwestern United States," *Arizona Anthropologist* 26 (2016): 53–75; Elaine Campbell, "Public Sphere as Assemblage: The Cultural Politics of Roadside Memorialization," *British Journal of Sociology* 64, no. 3 (2013): 526–47; Catherine Ann Collins and Alexandra Opie, "When Places Have Agency: Roadside Shrines as Traumascapes," *Continuum: Journal of Media & Cultural Studies* 24, no. 1 (2010): 107–18; Claire Corkill and Ray Moore, "'The Island of Blood': Death and Commemoration at the Isle of Man TT Races," *World Archaeology* 44, no. 2 (2012): 248–62; George E. Dickinson and Heath Hoffman. "Roadside Memorial Policies in the United States," *Mortality* 15, no. 2 (2010): 154–67; Margaret Gibson, "Death and Grief in the Landscape: Private Memorials in Public Space," *Cultural Studies Review* 17, no. 1 (2011): 146–61; Maida Owens, "Louisiana Roadside Memorials: Negotiating an Emergent Tradition," in *Spontaneous Shrines and the Public Memorialization of Death*, ed. Jack Santino (New York: Palgrave, 2006), 119–46; Sylvia Grider, "Roadside Crosses: Vestiges of Colonial Spain in Contemporary New Mexico," in *Descansos: The Sacred Landscape of New Mexico*, ed. Joan E. Alessi (Santa Fe, NM: Fresco Fine Art Publications, 2006), 11–28; Rebecca Kennerly, "Locating the Gap between Grace and Terror: Performative Research and Spectral Images of (and on) the Road," *FQS/Forum: Qualitative Research* 9, no. 2 (2008), n.p., located at http://www.qualitative-research.net/index.php/fqs/article/view/396; Una MacConville, "Roadsdie Memorials: Making Grief Visible," *Bereavement Care* 29, no. 3 (2010): 34–36; Richard Tay, "Drivers' Perceptions and Reactions to Roadside Memorials," *Accident Analysis & Prevention* 41, no. 4 (2009): 663–69; Richard Tay, Anthony Churchill, and Alexandre C. de Barros, "Effects of Roadside Memorials on Traffic Flow," *Accident Analysis and Prevention* 43 (2011): 483–86; and Karen Wells, "Melancholic Memorialisation: The Ethical Demands of Grievable Lives," in *Visuality/Materiality: Images, Objects and Practices*, ed. Gillian Rose and Divya P. Tolia-Kelly (Farnham: Ashgate, 2012), 153–69.

7 See John Belshaw and Diane Purvey, *Private Grief, Public Mourning: The Rise of the Roadside Shrine in British Columbia* (Vancouver: Anvil Press, 2009); Erik Cohen, "Roadside Memorials in Northeastern Thailand," *Omega: Journal of Death & Dying* 66, no. 4 (2012): 343–63; Mirjam Klaassens, Peter Groote, and Paulus P. P. Huigen, "Roadside Memorials from a Geographical Perspective," *Mortality* 14, no. 2 (2009): 187–220; Mirjam Klaassens, Peter D. Groote, and Frank VanClay, "Expressions of Private Mourning in Public Space: The Evolving Structure of Spontaneous and Permanent Roadside Memorials in the Netherlands," *Death Studies* 37, no. 2 (2013): 145–71; Rebekah Lee, "Death in Slow Motion: Funerals, Ritual Practice and Road Danger in South Africa," *African Studies* 71, no. 2 (2012): 195–211; Una Mac-Conville and Regina McQuillan, "Remembering the Dead: Roadside Memorials in

Ireland," *At the Interface / Probing the Boundaries* 58 (2009): 135–55; Olga Nespo-rova and Irina Stahl, "Roadside Memorials in the Czech Republic and Romania: Memory versus Religion in Two European Post-Communist Countries," *Mortality* 19, no. 1 (2014): 22–40; Anna Petersson, "Swedish *Offercast* and Recent Road-side Memorials," *Folklore* 120, no. 1 (2009): 75–91; and Anna Yudinka and Anna Sokolova, "Roadside Memorials in Contemporary Russia: Folk Origins and Global Trends," *Religion & Society in Central & Eastern Europe* 7, no. 1 (2014): 35–51.

8 Jennifer Clark, "Challenging Motoring Functionalism: Roadside Memorials, Heritage and History in Australia and New Zealand," *Journal of Transport History* 29, no. 1 (2008): 23–43; Jennifer Clark, ed., *Roadside Memorials: A Multidisciplinary Approach* (Armidale: Emu Press, 2007); Clark and Cheshire, "R.I.P."; Jennifer Clark and Majella Franzmann, "'A Father, a Son, My Only Daughter': Memorializing Road Trauma," *RoadWise* 13, no. 3 (2002): 4–10; and Clark and Franzmann, "Authority from Grief."

9 Clark and Franzmann, "Authority from Grief," 596. See also Lauren J. Breen and Moira O'Connor, "Acts of Resistance: Breaking the Silence of Grief Following Traffic Crash Fatalities," *Death Studies* 34, no. 1 (2010), 30–53.

10 Clark and Franzmann, "Authority from Grief," 594.

11 Clark and Franzmann, "A Father, a Son, My Only Daughter," 8.

12 Clark, "Challenging Motoring Functionalism," 33. See also Kurt Möser, "The Dark Side of 'Automobilism,' 1900–1930: Violence, War and the Motor Car," *Jour-nal of Transport History* 24, no. 2 (2003): 238–58.

13 See Campbell, "Public Sphere as Assemblage."

14 Ibid., 528.

15 Catherine Ann Collins and Alexandra Opie, "When Places Have Agency: Roadside Shrines as Traumascapes," *Continuum: Journal of Media & Cultural Stud-ies* 24, no. 1 (2010): 107–18, 110.

16 Clark and Franzmann, "Authority from Grief."

17 Patricia Ticineto Clough, ed., Introduction to *The Affective Turn: Theorizing the Social* (Durham, NC: Duke University Press, 2007), 1–33, 2.

18 See Sarah Ahmed, *The Cultural Politics of Emotion* (London: Routledge, 2004); and Thomas Rickert, *Ambient Rhetoric: The Attunement of Rhetorical Being* (Pittsburgh, PA: University of Pittsburgh Press, 2013).

19 Victor Buchli, "Introduction," in *The Material Culture Reader*, ed. Victor Buchli (Oxford: Berg, 2002), 1–22: 9. See especially, Ahmed, *Cultural Politics of Emotion*; Jane Bennett, *Vibrant Matter: A Political Ecology of Things* (Durham, NC: Duke University Press, 2010); Jill Bennett, *Empathic Vision: Affect, Trauma and Contemporary Art* (Palo Alto, CA: Stanford University Press, 2005); Elizabeth Edwards, Chris Gosden, and Ruth Phillips, eds., *Sensible Objects: Colonialism, Museums and Material Culture* (Oxford: Berg, 2006); Richard Grusin, ed., *The Nonhuman Turn* (Minneapolis: University of Minnesota Press, 2015); Carl Knappett, *Thinking through Material Culture: An Interdisciplinary Perspective* (Philadelphia: University of Pennsylvania Press, 2005); W. J. T. Mitchell, *What Do Pictures Want? The Lives and Loves of Images* (Chicago, IL: University of Chicago Press, 2005); Rickert, *Ambient Rhetoric*; Gillian Rose, *Visual Methodologies: An Introduction to*

Researching with Visual Materials, 4th edition (London: Sage, 2016); Gillian Rose and Divya Tolia-Kelly, eds., *Visuality/Materiality: Images, Objects and Practices* (Farnham: Ashgate, 2012); and Diana Taylor, *The Archive and the Repertoire: Performing Culture in the Americas* (Durham, NC: Duke University Press, 2003).

20 Rose and Tolia-Kelly, "Visuality/Materiality," 1, 3–4. See also Gillian Rose, *Doing Family Photography: The Domestic, the Public, and the Politics of Sentiment* (London: Ashgate, 2010).

21 Lawrence Grossberg, "Affect's Future: Rediscovering the Virtual in the Actual," in *The Affect Theory Reader*, ed. Melissa Gregg and Gregory Seigworth (Durham, NC: Duke University Press, 2010), 318.

22 W. J. T. Mitchell, "Introduction," in *Landscape and Power*, ed. W. J. T Mitchell (Chicago, IL: University of Chicago Press, 1994), 1.

23 Ibid., 1–2. See also Patricia L. Price, *Dry Land: Landscapes of Belonging and Exclusion* (Minneapolis: University of Minnesota Press, 2004); and D. W. Meinig, ed., *The Interpretation of Ordinary Landscapes: Geographical Essays* (Oxford: Oxford University Press, 1979).

24 John Urry, "The 'System' of Automobility," *Theory, Culture & Society* 21 (2004): 25–39. See also Mike Featherstone, "Automobilities: An Introduction," *Theory, Culture & Society* 21 (2004): 1–24; Mimi Sheller and John Urry, "The City and the Car," *International Journal of Urban and Regional Research* 24 (2000): 737–57; Mimi Sheller and John Urry, eds., *Mobile Technologies of the City* (London: Routledge, 2006); John Urry, *Sociology beyond Societies: Mobilities for the Twenty-First Century* (London: Routledge, 2000); John Urry, *Mobilities* (London: Polity Press, 2007); and Tim Cresswell, *On the Move: Mobility in the Modern Western World* (London: Routledge, 2006). For work specifically on automobility as a discourse and apparatus of neoliberal governmentality, see Jeremy Packer, *Mobility without Mayhem: Safety, Cars, and Citizenship* (Durham, NC: Duke University Press, 2008); and Cotten Seiler, *Republic of Drivers: A Cultural History of Automobility in America* (Chicago, IL: University of Chicago Press, 2008).

25 Michel Foucault, *The Archeology of Knowledge and the Discourse on Language*, trans A. M. Sheridan Smith (New York: Pantheon, 1972), 49.

26 See especially Kye Askins and Matej Blazek, "Feeling Our Way: Academia, Emotions and a Politics of Care," *Social & Cultural Geography* 18, no. 8 (2017): 1086–105; and Chantel Carr and Chris Gibson, "Animated Geographies of Making: Embodied Slow Scholarship for Participant Researchers of Maker Cultures and Material Work," *Geography Compass* 11 (2017): 1–10.

27 For case studies of this particular location, see David Kozak, "Dying Badly: Violent Death and Religious Change among the Tohono O'odham," *Omega: Journal of Death and Dying* 23 (1991): 2017–216; David Kozak and Camillus Lopez, "The Tohono O'odham Shrine Complex: Memorializing the Location of Violent Death," *New York Folklore* 17, no. 1–2 (1991): 10–20; and Rachael M. Byrd, "Rest in Place: Understanding Traumatic Death Along the Roadsides of the Southwestern United States," *Arizona Anthropologist* 26 (2016): 53–75.

28 See Hastings Donnan, "Material Identities: Fixing Ethnicity in the Irish Borderlands," *Identities: Global Studies in Culture and Power* 12 (2005): 96–99.

29 W. J. T. Mitchell, *Picture Theory* (Chicago, IL: University of Chicago Press, 1994).

30 See Darren Newbury, "Making Arguments with Images: Visual Scholarship and Academic Publishing," in *The Sage Handbook of Visual Research Methods*, ed. Eric Margolis and Luc Pauwels (London: Sage, 2011), 651–64; and Gillian Rose, *Visual Methodologies: An Introduction to Researching with Visual Materials*, 4th ed. (London: Sage, 2016), especially Chapter 12. Currently, a new form of cultural anthropology called "photo-enthnography" has produced a number of recent books with practices of using and balancing words and images similar to my own work. See especially Phillipe Bourgois and Jeff Schonberg, *Righteous Dope Fiend* (Berkeley: University of California Press, 2009); Jason De León, *The Land of Open Graves: Living and Dying on the Migrant Trail* (Berkeley: University of California Press, 2015); and Danny Hoffman, *Monrovian Modern* (Durham, NC: Duke University Press, 2017).

31 Lauren Berlant, *The Queen of America Goes to Washington City* (Durham, NC: Duke University Press, 1994), 4; and Rose, *Doing Family Photography*, 7.

Chapter 2

Trauma/Memory/Automobility

Early in Toni Morrison's 1988 novel *Beloved*, Sethe is talking to her daughter, Denver, about how it is hard for her to "believe" in time, and how it is confusing to think about the differences among her memory of places, her imagination of places, and the "real" places themselves.[1]

Sethe says, "I used to think it was my rememory. You know. Some things you forget. Other things you never do. But it's not. Places, places are still there. If a house burns down, it's gone, but the place—the picture of it—stays, and not just in my rememory, but out there, in the world. What I remember is a picture floating around out there outside my head. I mean, even if I didn't think it, even if I die, the picture of what I did, or knew, or saw is still out there. Right in the place where it happened."[2]

Denver asks if other people can see these "pictures," too, and Sethe replies, "Oh, yes, yes, yes. Someday you be walking down the road and you hear something or see something going on. So clear. And you think it's you thinking it up. A thought picture. But no. It's when you bump into a rememory that belongs to someone else."[3]

It's an evocative image to conjure here as we begin to understand witnessing the trauma of strangers in ordinary public landscapes: other people's traumatic memories taking the form of something you could physically and figuratively "bump into" as you move through the everyday world. What makes it evocative is its apparent simplicity but conceptual complexity. A rememory is not only a memory but a re-memory: a performative memory; an infinitely repeating memory; a memory of absence already once removed from the present that is nonetheless present; and an otherwise private memory that is always also present in public space. Significantly, Sethe sees such a rememory not only being located inside her own body or individual consciousness but also simultaneously "floating around" *and* anchored to a particular place "out there in the world," something somewhere where anyone—not just the people who already know the story behind it—could "bump into" it. In other words, Sethe thinks of her own rememories simultaneously as embodied and disembodied, individual and collective, invisible and visible, immaterial and material—as if rememory is a process, a force, and a thing all at once.[4]

Morrison's concept of rememory is certainly different from the dominant cultural idea of a ghost as something personified into a spectral body that lingers in a place, but it is also different from more theoretically complex conceptions of ghosts and haunting such as the one worked out in Avery Gordon's *Ghostly Matters* (2008). For Gordon, a haunting is not a supernatural force but a *social* force: Haunting is "a shared structure of feeling, a shared possession, a specific type of sociality"; it is "the sociality of living with ghosts, a sociality both tangible and tactile as well as ephemeral and imaginary."[5] The ghosts and the hauntings that Gordon analyzes are not material themselves; they perform absence through presence, paradoxically showing "a kind of visible invisibility: *I see you are not there.*"[6]

The crucial difference between a ghost and a rememory is in their materiality. To return to the central theme of this book, crash shrines are less like ghosts and more like *scars* on the landscape. They physically mark a traumatic disappearance and represent that absence in the public landscape. Like other public rememories, road trauma shrines unsettle and expose the incompleteness of the spaces they haunt by showing not only that *something is there* that shows that *something is missing* but also showing exactly *who* is missing, and that their absence is traumatic to those left behind. Their presence in the roadside evidences an absence, but, unlike ghosts, they have a material form themselves.

Shrines make themselves visible so that we can see that the object they "stand for" is not there, as if to say, *Look at me so that you can see that they are not here.* Unlike ghosts, which some people anthropomorphize and project upon and others deny altogether, shrines are both undeniably present and troublingly silent and still, while also seeming to be affectively dynamic. Shrines, like scars, are there for all to see and can have powerfully poignant effects on drivers and builders alike. Like scars, they can be concealed or ignored, but that does not mean that they are not still physically there. Finally, also like scars (and unlike ghosts), shrines do not transmogrify: shrines change over time and may even fade, but at every phase of their lives, shrines are always still shrines, until they are gone.

If not for roadside shrines, the potential for public memory of most fatal car crashes might evaporate, and places of automotive violence and trauma would be neither visible nor sensible even as *places*, much less recognized as *places of trauma* or *places of memory*. Like other unanchored rememories, they might still be "floating around" in the minds of those who experienced the trauma, and perhaps also "floating around," ghost-like, in the world. Then, they would be embedded incipiently into the landscape in intangible ways, as Sethe conceives them, where they could be either invoked by knowing story-tellers or sit awaiting the trained eye of the geographer, landscape historian, archeologist, or mystic to "read" them in the landscape. But encountering a shrine is encountering a road scar, not a free-floating apparition: shrines are a *concrete* instance of "bumping into a rememory that belongs to someone else." You don't have to know the trauma or even believe in shrines or even trauma itself to see them and feel something.

That's why, despite the fact that many people call them roadside memorials, roadside car crash shrines are not forms of memory, but rememory. They are involved in the process of memory, but do not themselves present or represent memory. Nor do they directly communicate the bodily trauma of crash victims or the trauma of insider witnesses. Instead, they perform trauma and memory through a series of materialized mirrors, each paradoxically reflecting an original trauma that itself cannot be figured directly but that is evident everywhere in roadside shrines. Encountering a shrine on the road is like all of the sudden coming upon a rememory: the evidence of an intimate, individual memory that belongs to someone else but is there for all to see. A crash shrine is like a broken bridge into another time, another reality. But it is also an unmistakably material thing, made from other undeniably thing-like things, all located in a radically particular place. You may not be able to use the bridge to get anywhere, but you can't deny that it is designed to be a bridge. You know that the memory is not yours, and it is not something you were looking for, but there it is, nonetheless, right in front of you, palpably real, and making a claim on you. And you can ignore the claim, but that doesn't make the claim go away.

A shrine thus makes a private memory of place into a social rememory of place: *Here. Here is where it happened.* You may not know what happened, but you will know that *something* happened, and *to someone, at this place.* A shrine anchors a "picture floating around there" and gives it a visual and material form in a unique spatial location. Like Sethe's example of a rememory of the house that burned down, a car crash shrine is a material reminder of trauma, just as a memorial to a house fire would be if it were placed on the site where a house once stood. The difference is that there is not a culturally recognizable form that signifies "fatal house fire occurred here," but there is a recognizable form that materially signifies "fatal car accident occurred here": a roadside car crash shrine.

Whatever else a shrine does, it always at least materializes rememory of trauma in a way that makes its fundamental claim inescapable and materially self-evident to even the most casual observer: *This is a place where someone was killed in or by an automobile.* Built by private citizens to remember private citizens and emplaced in public landscapes, roadside shrines thus ensure not only that road trauma rememories take a material form but also that strangers will "bump into a rememory that belongs to someone else" while they live their everyday lives being drivers, where roadside car crash shrines intrusively insert themselves into our lives to "accidentally" appear and remind us that the roadscape is a place not only of mobility but of that ultimate immobility: death.

Therefore, to track these processes and things animated by their forces is thus not only to follow the hauntings of ghosts but also to find them given material form in the landscape—and not just as individual rememories of trauma but also as collective ones. Indeed, in this context it is important to remember that Morrison's *Beloved* is not about the individual, social, and cultural manifestations of just any rememory but about the way that the traumatic memory of slavery is a *cultural trauma* that haunts the people and the nation of United States, whether people acknowledge it or not. In the novel and in the world, such a pervasive and often disavowed rememory permeates bodies, stories, pictures, objects, and places—emboldening some and driving others mad, but never simply "going away." It's like a scar on the skin of the culture.

Although the ongoing cultural trauma of automotive violence clearly has a qualitatively different character than the ongoing cultural trauma of slavery and racial segregation and violence—most importantly in its discriminatory power relations and its inherent entanglement with race and racism—there are important parallels. Both are massive, cultural traumas that are acknowledged and carried differentially within the culture. With both cultural traumas, intimates know the trauma only too well, but others have the privilege either to not to know the trauma exists or the privilege to think that it has nothing to do with them personally. The outcome is that while both are cultural traumas, some carry the weight of them more than others.

Roadside shrines embed their performance of traumatic affect and memory in the American roadside, forming a kind of dispersed but marked trauma rememory thoroughfare that has not (yet) been collected, mapped, and made visible *as* a structured phenomenon. One of my goals for the book is therefore to collect them—not only to make them visible as individual sites but also to make them comprehensible as a collectivizing cultural form. If we begin to collect these shrines, and see them working one by one to generalize their trauma, we can see that shrines give dispersed form to the cultural trauma of automobility embedded within contemporary American culture. Such a rememory includes the fatal bodily traumas experienced by not only the millions of people killed in and by cars over the past century, who will never run into a shrine even for themselves, but also the millions and millions more people like you and me who have been left behind to bump into them out on the road.

The purpose of the present chapter is to provide a theoretical framework for understanding how all of these different dynamics involved—trauma, memory, and automobility—work together in roadside shrines. Roadside shrines remember trauma through a distinctive visual/material/spatial form that, like trauma itself, intrudes upon the everyday spaces in which they are located, creating "accidental" encounters for drivers to recognize that the everyday traumas of automobility are not only individual traumas but also part of a collective cultural trauma. The best way to see this spatially defined dynamic operating is to place roadside shrines not only within the discourses of trauma, memory, and remembrance but also within the discourse of automobility. There, it becomes clear that when the family and friends of crash victims build shrines in the public right-of-way to help them work through their own traumas, these shrines also form a kind of vicarious trauma witnessed by other drivers, who are strangers to the people commemorated in shrines but definitely not strangers to the risks and rewards of automobility.

As crash shrines memorialize private individuals in public space, they embody a refusal to accept car crash deaths as collateral damage within automobility and an attempt to generalize individual losses into collective losses. And as they do, they make space for a quiet but palpable recognition that the everyday, ordinary traumas experienced within contemporary car culture are not only "their" losses but also "our" losses—that road trauma deaths are not an externalized by-product of automobility but are instead a central fact in and figure of the system itself.

TRAUMA SHRINES

Roadside car crash shrines are part of a wider worldwide phenomenon: something folklorist Jack Santino calls "spontaneous shrines."[7] With roots reaching deeply and widely through many different cultural traditions, shrines to people who die violently and suddenly in car accidents, murders, and political violence have proliferated since the 1990s. The early spontaneous shrines for the traumas of the Oklahoma City bombing, Princess Diana's death, and Columbine in the 1990s were amplified by the terrorist attacks of September 11 and throughout Europe in the early 2000s and the many mass shootings in the United States and beyond, establishing what is now an expected response to any public tragedy. Spontaneous shrines are located not in cemeteries or mausoleums where accident victims ultimately are "laid to rest" but within those *restless* spaces of everyday life where the unexpected deaths occurred—on roadsides, sidewalks, fences, transport stations, buildings, etc.

It is partially for this reason that Santino calls them "performative com-memoratives"; he argues that because they occur in public spaces, spontane-ous shrines are both *commemorative* (dedicated to sustaining the memory of individuals and events) and *performative* (meant to "make something hap-pen"—to materially transform the space of the event, the significance of an event, and anyone who interacts with the site). Santino emphasizes that spon-taneous shrines "insert and insist upon the presence of absent people"—they "place deceased individuals back into the fabric of society, into the middle of areas of commerce and travel, into everyday life as it is being lived."[8]

Similarly, in *Death, Memory, and Material Culture* (2001), Elizabeth Hal-lam and Jenny Hockey argue that material memory objects, sites, and practices used in more generally contemporary bereavement practices are "attempts to counter loss caused by death, making connections with the absent individuals and bringing them into the present."[9] Thinking of the material culture of grief and bereavement this way aligns with the dominant contemporary model of grief and bereavement, "continuing bonds," which emphasizes the need for continual working-through of grief and trauma instead of the older model of "getting over" grief.[10] For psychotherapists, the continuing bonds approach guides grief and trauma therapy, but for humanities and social science schol-ars studying grief and trauma, the approach helps explain the structure and function of both material and virtual memorial practices in the performance of trauma and grief through objects.

The way these many forms keep lost people alive socially is rooted in contemporary bereavement culture, where the cultural line between life and death also is becoming increasingly blurred. These shrines speak to and from "highly conflicted" contemporary attitudes toward death, and thus reflect and are structured within larger trends in death, dying, and bereavement practices.[11] Modernist practices of medicine, death, and burial that institutionalized living, dying, and burial made death not only more predictably ritualistic but also much less present and knowable in everyday life. People have begun resisting these practices, and have been develop-ing a set of counter-practices that have reasserted individual control over bereavement and burial or cremation. At the same time, cultural codes against showing grief in public have also shifted, emphasizing, if anything, now that people are expected and compelled to actively grieve in public, and are considered suspect if they do not. The outcome is that contempo-rary bereavement practices in the United States, the United Kingdom, and Europe are both more visible and more improvised, like other practices of today's diffused participatory culture rooted in neoliberal economic and social discourses, giving authority to (or leaving it up to, depending on your perspective) private individuals to decide the most appropriate ways of memorializing individual deaths.[12]

This is a point made also and more broadly about other contemporary memorial practices in the United States by Candi Cann in *Virtual Afterlives: Grieving the Dead in the Twenty-First Century* (2014). Cann argues that the current visibility of public grief for ordinary individuals is a response to living through the death-denying society of the twentieth century, which disenfranchised individual mourners and removed mourning practices from everyday life. As Cann writes, "Bereavement is no longer given public space in society or culture, which forces people to create and adopt alternative forms of mourning to help them navigate public space with their altered status as grieving individuals."[13] In the contemporary scene, "People need not only to grieve, but also to be publicly recognized as bereaved'" and cultural practices like roadside shrines, memorial T-shirts and car decals, tattoo memorials, and social media memorials enable that.[14] Similarly, Erika Doss argues that the growing presence of these public forms of mourning for private individuals suggests that "traditional forms of mourning no longer meet the needs of today's publics and prompt questions about what death, grief, and memory mean in the new millennium."[15] As we will see, the simultaneous individualization and "public-ization" of grief and mourning are also part of a larger ethical "duty to care" in contemporary neoliberal societies.

While many contemporary bereavement practices reflect this kind of "continuing bonds" approach to bereavement, practices associated with traumatic deaths particularly do. Indeed, one key factor in spontaneous shrines is the kinds of the deaths they commemorate. As Allen Haney, Christina Leimer, and Juliann Lowery established in the early 1990s, spontaneous shrines and the practices associated with them work to make sense of "unanticipated violent deaths of people who do not fit into categories of those we expect to die, who may be engaging in routine activities in which there is a reasonable expectation of safety."[16] Once focused mainly on car crashes, murders, and political violence, the practice since has spread to the mourning practices for many newer forms of unexpected deaths such as childhood cancers and still births, which have been made "surprising" in a sociocultural context where we "have gained such control over death that we now expect to die only of old age."[17]

As we will explore more fully throughout the book, such expectations of safety are not naturally occurring but are culturally produced within particular sociohistorical contexts. This is a point Marita Sturken also makes in *Tourists of History* (2007), where she argues that the traumatic "surprise" of events like the Oklahoma City bombing and the 9/11 terrorist attacks is in part attributable to the prevailing attitude toward history and public memory in the contemporary United States then.[18] Spontaneous shrines thus show just as much about a culture's expectations about safety and longevity as they do about how a culture handles the mourning of accidental deaths once they do happen.

They also reflect the culture's attitude toward innocence and guilt. The trauma materialized in spontaneous shrines is always figured as an external force, a force experienced as traumatic by someone who does not expect it. Thus, these deaths are usually coded as innocent, and the dead as the victims of trauma, even in the case of single-car accidents. As Karen Wells argues, "To be grievable is to be innocent."[19] As we will explore further later in this chapter, to be innocent in relation to automobility as a cultural discourse (and not only other drivers) is one of the main claims of a road trauma shrine.

When these attempts to counter traumatic loss through material cultural practices are placed in public landscapes, they make an implicit and some-times explicit demand on anyone who encounters them. As Avril Maddrell and James D. Sidaway argue, "Experiences of death, dying and mourning are mediated through the intersections of the body, culture, society and state, and often make a deep impression on *sense of self*, private and public identity, as well as *sense of place* in the built and natural environment."[20] Elizabeth Hallam and Jenny Hockey write that when a site of accidental death is actively performed and commemorated in public space, the site "material-izes memories"; "at stake is not necessarily a field of personal memories but, rather, the material relations through which past deaths are brought into the present."[21] For their lifespan at least, spontaneous shrines are dynamic spaces of intersubjective action, where material objects are located in public spaces and embedded in complex ordinary and extraordinary processes of being and becoming.

The contemporary landscape is saturated with these affective objects and spaces built by ordinary people to memorialize ordinary lives cut short by trauma. They are part of a larger trend in contemporary society toward the growing presence of vernacular (as opposed to institutional) memorial prac-tices in the everyday built environment and media environment. They are also part of the trend toward the spatial and temporal separation of memorial practices from the material disposal of bodies, what Leonie Kellaher and Ken Worpole call "cenotaphization."[22] Vernacular cenotaphs like roadside shrines, memory benches, and memorial trees planted in the everyday landscape work to performatively remember the dead by "anchoring memory to place" in the ordinary spaces where people lived and died, which constitutes not only an assertion of ongoing memory and social presence but also an implicit "form of resistance to the rapidity of change and standardization in the public realm."[23] Given the growing importance of social media as a "place" that people "inhabit" socially, it is no surprise that the process of cenotaphization is present in virtual memorial forms online as well as material ones.[24]

These kinds of cenotaphs go by a number of other names in contemporary public discourse: spontaneous memorials, grassroots memorials, temporary memorials, vernacular memorials, impromptu memorials, and makeshift memorials. Of the existing terms, I prefer the term "spontaneous shrines" because it best combines what I consider to be the two criterial elements of the practice: the word "spontaneous" captures the apparently self-generating, always-already produced quality of the form as it seems to appear automatically after such a tragedy. And the word "shrine" captures the dynamic, sacralized, intersubjective, and performative quality of the form, which redefines an everyday space and otherwise ordinary objects into an excessive, affectively "heavy" and participatory space the way other shrines do. Although some may resist the term "shrine" because shrines are often associated with religious or spiritual beliefs and practices, I see all kinds of shrines being more generally about asserting and materializing a kind of sacralizing of lost, out of place, or hard-to-reach objects, regardless of the religious and cultural tradition within which they are performed.

However, what all of these names for the cultural form lack is any direct reference to their original impetus—the trauma of experiencing the unexpected and violent death of a friend or family member. All of these shrines mediate the trauma of sudden, violent death and grief. They help people process troubled and troubling deaths. More specifically, car crash shrines would not exist if the deaths they mark were not traumatic—traumatic in terms not only of the body trauma suffered by those killed in and by cars but also traumatic for those left behind to suddenly make sense of their unexpected loss, made all the more unexpected by the fact that they were killed driving or riding in an automobile, one of the most ordinary activities in contemporary American culture. Finally, as I have already begun arguing in relation to my response to the photograph of the boy in braces at that California shrine where we started the book, roadside shrines both materialize and produce trauma for strangers was well.

For that reason, I will use the term "trauma shrines" throughout the book when I am referring to the general practice of building a shrine to mediate a sudden, violent, and unexpected death, and "road trauma shrines" or "roadside car crash shrines" to refer to those trauma shrines pertaining to sites for automotive crash deaths. These terms reinscribe violent trauma into the sites as a central dynamic and central material reality in more ways than one. As we will see, trauma shrines are used by intimates to work through the process of mourning traumatic deaths, and as they perform this function, they also evidence that trauma to strangers through visual, material, and spatial means, creating a potential not only for secondary trauma for them as well but also a certain horizontal affiliation with others who witness the shrines alongside them in public space, which is the central dynamic this book analyzes.

It is also important to reinscribe trauma into our conceptions of these shrines because it helps explain why they are becoming more prevalent in the contemporary cultural environment. Faced with multiple forms of trauma at

personal, interpersonal, local, national, and global scales, living in contemporary America is not only about living in trauma but also about living with the
consciousness of living in trauma. Indeed, Nancy Miller and Jason Tougaw
have argued that we are now living in "the age of trauma," which is apparent not only in the traumas people are encountering but also in the extensive
public discourse about trauma and the growing scholarly attention given to
the topic as well.[25] The result is a culture where trauma is not only experienced by many but also claimed and mobilized as a public affect that forms
the basis of truth claims in cultural discourse, where people who have experienced trauma are said to "know trauma" but not be able to "show trauma,"
and where others are asked to treat that embodied knowledge of trauma in
certain ways even though it can neither be known by nor communicated to
others directly.

ENGAGING TRAUMA

Given the importance of trauma to the culture and to trauma shrines and road
trauma shrines, we need to better understand trauma before we move forward.
The dominant model of trauma in the contemporary interdisciplinary field
of trauma studies emphasizes dissociation and the belatedness of trauma
memory.[26] In her foundational work on trauma within literary studies in the
1990s, Cathy Caruth argues that traumatic memory is distinctive not for the
content or character of a particular event but for "the structure of its experience or reception: the event is not assimilated or experienced at the time, but
only belatedly, in its repeated possession of the one who experiences it."[27]
Trauma separates the self from the conscious cognitive experience of and
thus the memory of a traumatic event, which makes traumatic memory function outside narrative, living in the affective realm, where it is primarily experienced in belated, latent, intrusive repetition. Consequently, as Ann Kaplan
writes, because trauma does not produce narrative memory and "because the
traumatic experience has not been given meaning, the subject is continually
haunted by it in dreams, flashbacks, and hallucinations."[28]

Scholars theorize trauma's belatedness as an effect of the experience of
trauma itself: during a traumatic experience, the mind protects the body
by denying its ability to fully perceive the trauma as it occurs. But that is
exactly why the traumatized are said to "act out" their trauma instead of com-
prehend it. Roger Luckhurst argues that the traumatic subject is "dispersed
'horizontally' in various forms of dissociation. It cannot remember itself to
itself; it has no cohesive narrative" with which to make sense of its traumatic
experiences.[29] Similarly, Ulrich Baer characterizes trauma as "a disorder of
memory and time" that "imposes itself outside the grasp of our cognition."[30]

The goal of individual trauma therapy is then to consciously "work through" trauma to bridge that gap so that the subject can not only process the experience and integrate it into identity but also communicate that memory to others—all while knowing that such an endeavor is never-ending, frustrating, and likely to fail.

This understanding of trauma as a belated, unrepresentable, uncontained, and intense experience originated in psychoanalytic theory and practice, where it was first applied in individual psychotherapy. But trauma studies scholars also have applied this model to understand collective forms of trauma such as war, genocide, forced migration, and natural disasters as well, developing a body of work particularly concerned with the role of *collective trauma* in both the need for and struggle over collective memory.[31] This work has identified collective traumatic effects at multiple cultural scales, from small groups to subcultures to whole cultures to transcultural diasporas and even, more problematically, to all of humanity.

For the purposes of understanding how trauma works at road trauma shrines, the most pertinent work here is on mediated and vicarious trauma. For example, scholars analyzing the news media coverage of events like September 11 have argued that the *mediated* collective experience of trauma takes similar forms as it is incorporated into cultural memory. For example, Alison Landsberg characterizes the shared memory of mediated events as a form of "prosthetic memory"—a concept similar to rememory in that it is a memory you can have of someone else's memory that is only available to you through some mediation.[32] Similarly, E. Ann Kaplan applies the terms "secondary trauma" and "vicarious trauma"—terms originally used by psychotherapists to describe the traumas therapists experience as they help their clients work through trauma—to contemporary media audiences' experiences of trauma through television, film, and photojournalism.[33]

Although personal and cultural traumas, just like personal and collective memories, are not equivalent, cultural trauma has a similar dynamic: overwhelming in its unfolding, the recognition and communication of cultural trauma is also belated, where witnesses work through their own trauma while also addressing others to materialize the truth of their trauma. The important point to emphasize here is that trauma may escape conscious perception and resist integration within *cultural* bodies as well as *individual* bodies. The question then is: if a group of people experiences trauma together—at whatever scale—does it produce a similar dissociative process that shapes how that trauma is perceived, reencountered, and communicated within and by the group? Does the memory of large-scale cultural trauma take the form of "acting-out" though repetitive and intrusive but belated "dreams, flashbacks, and hallucinations" at the level of the culture as well? And if it does, where and how are these intrusive materializations of such a cultural trauma located so that the culture might learn to "remember itself to itself," as Roger Luck-hurst says?

However, as a number of critics have pointed out, the larger the scale of the collective becomes, the more problematic the analogy between individual trauma and collective trauma becomes.[34] This is for three reasons. One is that people using the trauma concept at larger scales often either forget or ignore that they are using a metaphor: that individual bodies and collective bodies are not equivalent, so to argue that collectives experience trauma in the same way as individuals do is always to make an argument, not an observation. The second reason is that the experience of trauma is highly variable, mean-ing that different people experiencing the same event will experience differ-ent magnitudes of trauma, with some not experiencing it as traumatic at all. Trauma is a relationally defined feature of a response to an event, not intrinsic to any particular event. That means that we should be careful not to essential-ize trauma or to simply assume it is occurring without doing a fine-grained analysis when it appears on the scene. And finally, the third reason is that stretching the trauma concept too far loses the distinctive nature of trauma as a force involved in how collectives work through and remember traumatic experiences. If trauma interrupts and frustrates both memory and communi-cation of that memory, and if trauma authorizes claims to knowledge beyond critique, then naming something as trauma can become an end in itself that closes off analysis instead of opening it up.

While some scholars refuse to consider collective trauma as a concept at all and others use it indiscriminately, the better course is to recognize the limitations of the analogy between individual and collective traumas and map it out carefully while doing empirical research. That indeed is my approach to understanding the traumas performed through road trauma shrines by the dif-ferent collectives who work through them: from the intimates who build and maintain them to the strangers driving by them, and even to people like you reading about them in books. You will see me map out these different types of collective trauma most extensively with regard to road trauma shrines in chapters 6 and 7, but here I want to identify some of their key features.

The most important theoretical distinction that needs to be made in ana-lyzing collective trauma is in defining the nature of the collective itself. For instance, in their apt critique of the "imperialistic" and "cavalier metaphorical application of trauma" within contemporary trauma studies, Emily Keightley and Michael Pickering write, "The extrapolation of trauma to large groups of people on the basis of secondary or tertiary encounters with violent or disrup-tive pasts hugely dilutes the experience of victims and survivors, and in doing so empathy, as a response premised in the maintenance of critical distance, is transformed into a position of identification."[35] The problem with this char-acterization is that it assumes that collective trauma operates only vertically instead of horizontally, where trauma is conceived of as being centered on the response to an event that is witnessed at different scales and that the collective includes everyone at every scale. If we center the concept of cultural trauma on the experiencing of a traumatic event or set of events, then yes, identification-focused secondary and tertiary witnessing can be colonizing and diluting because it is claiming to know something that it cannot. But what if we consider that there are collectives formed *across and among* the separately scaled groups: primary witnesses, secondary witnesses, and tertiary witnesses (and so on)? What do secondary witnesses share as a collective in relation to the trauma they experience secondarily alongside other secondary witnesses, and how is it similar to and different than what a collective of primary or tertiary witnesses experiences (and so on)?

Further, where most analyses of cultural trauma assume a defined precipitating event experienced either in person or through mediation or are the result of differential power relations that systematically produce perpetrators and victims, the cultural trauma caused by extensive and ongoing individual road trauma does not work the same way. Currently, it is registered as a statistical collective more than a cultural collective—a *collected memory* instead of a *collective memory*.[36] Embedded as they are in someone else's territory, radically tied to a particular location on the roadside, unable to police their own boundaries, heavily regulated and actually illegal in most states, and unable to make themselves officially known as public memories, each shrine fights its way into collective consciousness and collective memory individually, one by one.

That is the main reason road trauma does not (yet) feel like a cultural trauma. Indeed, if we use Jeffrey Alexander's definition of cultural trauma, that makes sense. Alexander argues that cultural trauma is the result of *identity trauma* within a group: "Trauma is not the result of a group experiencing pain. It is the result of this acute discomfort entering into the core of the collectivity's sense of its own identity."[37] And that's again where automobility comes back into play. The culture is so implicitly committed to automobility that it has not (yet) become significantly upset by the collateral damages it produces to feel it as an identity trauma.

The people undergoing road trauma most directly—the people building, rebuilding, and visiting road trauma shrines—are undergoing identity trauma, and they are trying to show the rest of us that we are disavowing our own identity traumas. The work is not taking the form of a social movement, however, with clear political goals, representative leaders, or even a clear sense of being in a collective. Though often disavowed and definitely not settled, significant social movements have ensured that the cultural traumas of large-scale ongoing mass traumas, such as the atrocities of the Holocaust, slavery and Jim Crow, the genocide of Native Americans, and

Euro-American colonialism, are now registered as cultural and even global traumas. Significantly, all of these traumas unfolded within a clear victim/ perpetrator discourse not present in car crashes—with the exception of social movements to end drunk driving. Instead, the collective forms of road trauma are taking the form of a dispersed collection of individual road trauma shrines, an aggregate memorial trying to remind us about something we may have never experienced ourselves directly but nonetheless know exists: the fact that many, many people have died, are dying, and will die in car crashes.

My work on shrines seeks to bring together the two main strands within studies of cultural trauma—the study of isolated individual experiences of large-scale trauma and the study of vicarious witnessing of distant or virtu- alized trauma—to theorize roadside shrines as isolated, individual material forms of vicarious trauma memories of an otherwise large-scale trauma dispersed throughout the material and cultural landscape. With large-scale traumatic events such as 9/11, those efforts can be centralized in one par- ticular space or a small number of spaces. With car crashes, however, they are extraordinarily diffused, so the testimony *and* the witnessing happen one trauma at a time, one space at a time, one witness at a time.

ROAD TRAUMA SHRINES AND CULTURAL TRAUMA

A single car crash is immensely traumatic to the people directly involved, and friends and family members who build shrines create a shared space for mourning through the shrine. But what about for the larger public, for those of us "just looking" as we drive by? How is road trauma shared publicly, and what kind of collective is formed through that sharing? How is it similar to and different from other forms of traumatic collective memory? When individual traumatic memories of road deaths are shared in public, the act of sharing them opens up a potential space for connection through the experi- ence of vicarious trauma through secondary witnessing. In this, they are simi- lar to other "public" traumas like warfare, natural disasters, public transport crashes, mass and serial murders, and genocidal atrocities, which often are simultaneously witnessed by many people directly and witnessed by even more people through media.

For instance, Roger Luckhurst argues that when traumatic subjects form a collective, it is a potentially intense but always also contingent, fragile, and effervescent public, built on what he calls "temporary communalities."[38] Indeed, the collectivizing effects of public trauma, grief, and mourning are central to contemporary American citizenship today. As Edward Linenthal argues in his book about the Oklahoma City bombing, *The Unfinished Bomb- ing* (2003), public-mediated mourning is one of the few forces that creates at least a momentary experience of unity for the contemporary diverse nation: "being 'together' with millions of others through expressions of mourning bypasses or transcends the many ways in which people are divided—by religion, by ideology, by class, by region, by race, by gender."[39]

However, the trauma represented by road trauma shrines is both unlike the sustained, collective, and catastrophic trauma of war, genocide, terrorism, or forced migration *and* unlike the equally dispersed but more distanced screened experience of witnessing someone else's trauma through media technologies as a dispersed audience. The only way to know the affective intensity of the embodied trauma of a fatal car crash is to die in the car. And the only way to know the trauma of losing someone you know in a car crash is to experience that directly yourself. No one would wish that on anyone, but they might want someone to know what it feels like, or at least want someone to know that they have experienced it and now are living with it. And then there might also be reverberating effects at a much larger scale once those performances of traumatic grief are spread across the shared landscapes of automobility. That is because road trauma shrines confront people with the demonstration of a trauma not on an electronic screen or within a museum building and behind velvet ropes, but out there on the roads and streets of the automotive landscape where car accidents themselves occur. There, they address anonymous witnesses and demand witnessing from a particular spatial and discursive location: through the windshield of a moving automobile—an automobile in most cases driven by a person doing the same thing the person who died was doing when they died. That means that the only way to experience one in person is to put yourself in a situation where you risk becoming part of the display.

Because they are experienced primarily by drivers while they are ensconced in the familiar embodied environment of a car interior moving at speed through the landscape, shrines serve to map crash traumas directly to the everyday experience of driving. That is, roadside shrines are emplaced in such a way as to make accidents visible to people in cars going about their everyday lives as drivers and passengers, where people are forced to encounter evidence of the "negative" fatal accident while performing the "positive" of autonomous mobility.[40] There, shrines interpellate visitors through complex dynamics of mirroring and distancing that *accidentally* expose drivers to "the car accident" as a physical and discursive phenomenon to viewers and expose the intrinsic, embodied connections between those viewers and the accident.

Once placed in an everyday space, crash shrines mark a place where trauma disrupted the flow of everyday life for the victim and everyone who knew them. Like large-scale, collective traumas such as 9/11, car crashes are located in social space. Because they are placed in shared space, this trauma is potentially always simultaneously private and public. But unlike large-scale traumas, road shrines mark dispersed, privately managed traumas that are felt deeply by individuals but rarely the whole community, much less the nation. That is why I think of them as *ordinary traumas*: they are ordinary both because they occur to ordinary people as part of the ordinary mobilities of everyday life and because they are rarely recuperated into some larger collective that can make them seem extraordinary. And because these ordinary traumas happen *all over the place, all the time*, they don't register as a mass trauma the same way that event-based mass traumas do.

This makes the ordinary traumas of automobility similar to the kinds of everyday traumas and ordinary publics engaged in the work of Ann Cvetkovich and Kathleen Stewart on public affect and affiliation. Cvetkovich has collected an analysis of lesbian "sites of trauma" into "an archive of feelings," which has allowed her to identify and explore "a sense of trauma connected to the textures of everyday experience" for a culturally defined collective—in this case particularly a cultural minority—where "affective experience can provide the basis for new cultures."[41] Cvetkovich shows how trauma texts—*and* the act of collecting and curating dispersed trauma texts—can create temporary affective affiliations that can be used to both break through existing collectives and form new collectives. Likewise, Kathleen Stewart collects everyday acts that seek to create at least a temporary "we-feeling." Such an "ordinary affect . . . permeates politics of all kinds with the demand that some kind of intimate public of onlookers recognize something in a space of shared impact. If only for a minute."[42]

A roadside shrine is just such a momentary "space of shared impact," where drivers speeding by a shrine are presented with intrusive cultural flashbacks of vicarious traumas that pertain directly to the activity they are presently embarked upon: using a private vehicle to move through the public spaces of automobility. Thus, shrines form a dispersed material "archive" of traumatic experiences within the public right-of-way that, as Stewart describes other momentary publics, "demand that some kind of intimate public of onlookers recognize something."[43] Shrines create a place where a "little world comes into view," as Stewart would characterize it, and a " 'we' snaps into blurry focus," where those in the "we" are met with "a sense of shock or relief at being 'in' something with others."[44]

Like the affiliations focused on ordinary trauma that Cvetkovich and Stewart have studied, shrines are mediations of ordinary trauma that themselves take the form of traumatic memory—intrusive, affective, visual, material—and do so not just for the people who knew and loved those who died at the site but also for the rest of us strangers who drive past them. When drivers encounter them, they intrude on our attention like a flashback or rememory belonging to someone else. If the spatial anchorage of all trauma shrines is inseparable from what they do and how they do it, the space where road trauma shrines exist is particularly important. Instead of being contained in a finite time and space where it can be celebrated, suppressed, managed, or ignored, however, affective traumatic rememory intrudes upon everyday consciousness, demanding an embodied experience of someone else's trauma—right there, right then, all over again. The something that precarious "we" is "in" is the shared experience of witnessing someone else's road trauma alongside other drivers doing the same.

A road trauma shrine is not only a materialization of the trauma of the individuals involved in that specific shrine but also a materialization of the larger *cultural* trauma associated with cars and car culture—the now century-long cumulative trauma of unassimilated, abject, but also somehow mundane deaths. Roadside car crash shrines can be consciously or unconsciously ignored, but once they do register with drivers, they work to represent and perform vicarious trauma—an intrusive, accidental, and repetitive reminder of the trauma of others that simultaneously *speaks to* and *speaks of* automobility as it *speaks out of* automobility.

Here I am drawing on Jill Bennett's distinction between affective memory and narrative memory. Where narrative memory works in and through linguistic representation to be *about* a memory, affective memory is more like a flashback or a rememory, where "affective experience is not simply referenced, but activated or staged in some sense" by a process of "registering and producing affect" in an experiential way, which produces "not so much a *speaking of* but *speaking out of* a particular memory or experience."[45] Bennett generates this distinction to help her make sense of contemporary art practices and work that evoke trauma not through representing trauma directly but through "a certain affective dynamic internal to the work" that *reproduces* the dynamic of trauma more than *represents* it.[46] Thus, like road trauma shrines, the art Bennett analyzes is not *about* trauma; it *does* trauma. Bennett argues that such work "might not readily conform to the logic of representation";[47] instead, it is structured in affect, creating encounters between the artist, the artworks, and viewers that perform trauma and memory through material and visual means. Such art is thus "*transactive* rather than *communicative*";[48] it does not communicate existing truths to a stable and separate audience but is a location of encounter. In this moment of contact between bodies that feel traumatic memories, even strangers potentially can feel another's pain as their own, not in an act of colonization of the Other or even of projection of sameness, but just the opposite: feeling another's pain as a wound or scar that ruptures the subject/object split instead of as the distanced pain of a contained Other. It is "the point at which one both feels and knows feeling to be the property of an other."[49]

As we will see in much more detail later, like the art practices that Bennett has studied, road trauma shrines do not *represent* or *speak about* trauma; they *reproduce* and *do* trauma. They do not literally represent the trauma directly so that it can be communicated directly to strangers, which would assert both the potential knowability and conveyance of trauma. Instead, they perform the trauma that the shrine builders know, but do not translate or represent that trauma directly to strangers. Instead, they *speak out of* trauma to create a transactive space of affective encounter between intimates and strangers that has the potential to show both that each individual road death is not only an individual trauma but also part of a cultural trauma. Built to negotiate trauma, that is, they also take *the form of* trauma in the roadway: an intrusive counterbalance to the disavowed knowledge that cars kill. They materialize a belated understanding of something we all have been experiencing but have not consciously worked through as a culture.

Shrines are an intrusive form of traumatic memory that finds you where you are, asking you for something, but not able to be clear about saying what they want. Individual shrines shout at passing drivers, but they also stammer. They communicate that they know something that cannot be communicated, at least not directly through anything approximating a language. This is a crucial point, because this is not a book that interprets and translates shrines, that "reads" shrines like a text to reveal their meaning. It is a book that attunes to the subtle and inarticulate ways they *present* (instead of *represent*) and *perform* trauma.

Moreover, when you look at shrines together, as I have done in this book, they become more articulate. When you drive by a shrine and notice it, you are brought into its melancholic web of transference, where you may or may not feel what Kathleen Stewart calls a "shared impact." And when you *keep driving* by that shrine, and keep driving by even more of them, the effects and affects can multiply. Encountered in the spaces of individualized mobility, a shrine is a heavy anchor that asks a culture structured in and performed through forgetting, innocence, speed, and mobility to recognize the constraints and losses encoded into and negated by that speedy drive into the future. Shrines ask drivers to slow down—not just in the sense of cautiously slowing down their cars because they are negotiating a section of road that has already claimed at least one fellow automobilist but also to slow down and reflect on their participation in automobility itself. They ask drivers to reflect on their belonging to a collective that never quite feels like a collective, and to take account of their ethical responsibility for recognizing and remembering the collective loss that shrines perform.

HOW ROAD TRAUMA SHRINES SPEAK FROM
AND TO AUTOMOBILITY

Every year in the United States, around 35,000 people die and over 2 million people are injured in some of the more than 6 million car crashes reported to law enforcement agencies. In recent decades, the numbers have mostly trended downward from a peak of nearly 55,000 reached in 1973 when there were far fewer drivers and cars on the road in the United States. The overall numbers trended down until 2011, when they were 32,479, their lowest point since 1949, but are currently on an upward trend started in 2012, with the most recent reported year of 2016 seeing 37,461 traffic fatalities in the United States.[50] Calculating not only the economic costs of crashes but also the "quality of life lost" caused by car crashes, including "lost market and household productivity," the National Highway Transportation Safety Administration has estimated that these crashes have a "comprehensive cost to society" of $836 billion a year.[51]

And even with a significant decrease in the scale of these deaths over the past few generations, especially given that the number of drivers has grown considerably in the same time, these are enormous aggregate numbers. They add up not only to significant measured economic costs but also to the loss of a total of more than 3 million *individuals* killed in traffic fatalities in the United States since the early twentieth century. These millions of road deaths in the United States over the past century have traumatized the culture as a whole.[52] They amount to a cultural trauma that is embedded within the larger American cultural experience and identity. I'm not just talking about the individuals and groups who have absorbed the trauma of these road deaths directly but the culture at its largest macro scale, and everyone in it.

And yet, because they happen at isolated times and places, fatal car crashes never quite cohere into a recognizable *collective* cultural trauma the way war, political violence, and natural disasters do. As Gregory Ulmer suggests, one reason why road deaths are not memorialized in mass form in a kind of general car crash memorial at the national scale is that they do not *occur as* mass deaths. Michael Warner makes a similar point in *Publics and Counterpublics* (2002), arguing that only mass traumas generate mass mourning, leaving the work of making diffused or minority traumas public to a "counterpublic" that then must address a larger public to demand recognition.[53] Road deaths are dispersed enough in space and time to never coalesce into a feeling of mass trauma, which leads to a kind of structural forgetting of the cultural trauma of road deaths. That is, the culture knows the trauma of automobility, but does not know that it knows it. Instead of working-through it, it either disavows it or acts it out.

Regardless, the effects—and definitely the affects—remain. The unspoken, invisible past traumas of car crashes haunt American cultural memory as much as they haunt the landscape of roadside America and the people who drive through it. This haunting takes many forms—in personal memories of particular crashes experienced or witnessed, in the many transgenerational stories of ancestors who perished on the road, in a cultural obsession with notorious celebrity car crash deaths, and in the seemingly endless spectacle of cars crashing into one another on our media screens.[54] Of course, the car crash also is encoded into the very form of automobiles and the infrastructure that they inhabit—inscribed into cars in the form of bumpers, seat belts, air bags, etc., and into the roadscape in the form of crash barriers and traffic warning signage, etc.

But while a car's bumpers and its chassis are designed to absorb and distribute the energy of a crash, and crash barriers are designed to do the same, there is no institutionally legitimated cultural technology to serve that purpose for the people involved in car crashes or the survivors who mourn them. There is no national car crash memorial, no centralized large-scale cultural practice of memorializing car crash deaths, no corresponding social movement—and thus no means of generalizing the trauma, no way of "translating the energy" of a crash into a politically significant process of public memory. Instead, the cultural work of remembering people killed in and by cars is left to individuals constructing roadside car crash shrines on their own—sometimes individuals working within state-sponsored memorial programs or with advocacy groups like Mothers Against Drunk Driving (MADD), but more likely individuals acting on their own. Ubiquitous but dispersed, roadside shrines do not cohere in time or space to seem like anything more than a statistical collective instead of a cultural or social collective; they seem to form more of an aggregate than a collective. The statistics slowly pile up, making a significant number in the aggregate, but individual road deaths create only localized intimate publics.

The best way to understand the distinctive character of the traumas that road trauma shrines figure is thus to understand their place within the discourse of automobility, which produces an additional layer of structural forgetting about the everyday trauma of living with cars in the United States. This involves seeing car crash shrines—and the deaths and crashes they memorialize—as part of a vast cultural apparatus, what I have been calling automobility. Like the automobile it derives its name from, automobility as a term is a combination of *autonomy* and *mobility* that refers not only to the "self-propelled" nature of car engines but also to a self-propelled discursive system of producing and regulating automobile-bound performances of autonomy through mobility. A vibrant interdisciplinary field called automobility studies, a subset of mobility studies, has developed in recent years to study automobility.[55]

Like automobiles themselves, which Paul Gilroy describes as "probably the most destructive and seductive commodities around us," automobility as a cultural discourse is complex and contradictory.[56] But where the automobile is both an object and a figure, automobility is a process and a system of objects and subjects at the same time, increasing both complexities and contradictions. Mike Featherstone writes that automobility "speaks to powerful cultural dreams of adventure and freedom: the capacity to go anywhere, to move and dwell without asking permission, the self-directed life free from the surveillance of the authorities."[57] But Featherstone and others also recognize that these dreams of freedom are embedded in discourses, regimes, systems, and cultures that both enable and control the dreams and the subjects who drive under their influence.[58] Where being a passenger obviously involves giving up autonomy to ensure mobility, driving a car has a more paradoxical relationship to autonomy. Driving is paradoxical because it is a practice where expert systems permeate every aspect, rendering drivers subject to enormous mobilizations of expert knowledges, control mechanisms, etc., but it simultaneously also must be performed and negotiated by millions of private individuals who perform "both autonomous self-direction *and* acquiescence," one at a time, together, at speed.[59]

Like other scholars of automobility, such as Cotten Seiler and Jeremy Packer, I have turned toward Foucault's work on discourse, apparatus, governmentality, and technologies of the self to understand how the system of automobility not only organizes spaces and objects but also produces subjects who will perform automobility within and in relation to the always discursively situated spaces and objects within the apparatus (and, of course, in relation to other subjects).[60] For instance, bringing together Foucault's work on governmentality and Deleuze and Guattari's work on smooth and striated spaces and the control society, Packer sees contemporary automobility being fully embedded in a neoliberal political economy, where "the goal of governing is not to simply guard against too much freedom, but to produce the type of freedom that accords with the expansive demands of culture and economy. Governing at a distance across striated space takes the place of direct control."[61] And out there on the road, and on city streets, where each driver must perform automobility alongside other neoliberal subjects, governmentality— what Foucault calls the "conduct of conduct"—must be continuous, flexible, and internalized if it is to function at all.

To drive in this world, "Mobile subjects . . . must be highly disciplined, because they are not under continual surveillance."[62] Certainly, as Packer also points out, traffic policing is extensive, but the system depends on a panoptic internalization of drivers who know they *could be* monitored and thus *potentially* under surveillance at any moment. However, traffic enforcement operates under the repressive hypothesis but does so in a system of mobile "government at a distance" instead of through enclosure. This makes it different from more sedentary, strategically emplaced institutional spaces like prisons and airports where institutions and organizations can force people to conform to institutional control by channeling them through checkpoints and other more invasive policing regimes. On the streets and highways of America, the police must generate "voluntary compliance" through self-discipline not only by being visibly present often and everywhere for drivers to internalize the surveillance but also by actively giving citations to somebody, sometimes. Packer argues that this alone is not enough to generate compliance, though, for the "link between panopticism and the internalization of the gaze depends, at least in this instance, upon the internalization of a particular discourse, safety."[63] That is, the discourse of automobility produces not only technologies and policing practices designed to facilitate safe movement through roadspaces but also produces drivers as *subjects* and members of a *public* who must recognize that they themselves ultimately are responsible for their own safety within that infrastructure.

Like any other discursive formation or apparatus, then, automobility produces not only a larger collective—the motoring public—but also a group of subjects who are both subjected to the apparatus and given agency within the apparatus, who must govern themselves alongside others within that collective. If we call the collective produced by automobility the "motoring public," then we might call an individual subject in the system the "driving subject" or the "automotive subject." As we will explore much more fully in chapter 6, where we engage the process of interpellation extensively, the motoring public paradoxically is called into being as it is addressed but also addressed as if it already exists. That is, throughout the culture but especially on the road, individuals are constantly addressed as if they are always already members of the motoring public: in official signage, in commercial billboards, in commercial architecture, and even in the structure of lane markings on the road surface itself.

Most institutional forms of addressing the motoring public invoke "public safety" as the overriding value here: *our job is to keep you safe; we do what we do to keep you safe; you need to comply with what we say to keep yourself and others safe; otherwise, you are at fault.* These appeals are structured within what Joseph Gusfield has called the "crashworthiness approach" to road safety, where accidents are factored into the car manufacturing industry and road safety system, and where the focus is on developing and implementing new technologies to minimize the impact of such accidents, not removing accidents themselves from the system.[64] The system of automobility dominant in the United States today would best be characterized as what Beate Elvebakk calls "mitigated liberalism," where drivers are relatively autonomous as long as they internalize and follow the basic rules of engagement within the system, and where car manufacturers and road safety institutions negotiate their own power interests and publics to establish the conditions necessary for drivers to be safe from and safe in crashes. In this model, Elvebakk writes, "Drivers [are] held accountable for their actions, but they, their vehicles and the system in which they interact" are "being made continually safer through public control and intervention."[65]

If, as Cotton Seiler argues, driving is the practice that performs one's participation in a liberal democracy, then it seems almost inevitable that any one driver's autonomy is on a collision course with other autonomous driver-citizens following their own prerogatives. Thus to drive a car on roads where driver choice is not entirely constrained (the epitome of Foucault's self-disciplinary subjectification) and where the implications of another driver's lack of self-governance (and, of course, one's own) are particularly grave, driver-subjects constantly must subject themselves to proper codes of conduct while simultaneously hoping that other drivers are doing the same or at least watching out for those who are not.

Drivers must place their trust in other drivers as well as in road safety experts to design the roads, automobiles, and traffic management systems as well as they can. But to be a driver is also to take responsibility for nego-tiating those systems as effectively and ethically as possible while simul-taneously hoping for the same among other drivers. Of course, as the term "defensive driving" illustrates, drivers are also expected to hedge their bets. Drivers need to believe the system is safe to be able to perform it. The state needs drivers to act as if the system is safe, as if they can trust other drivers to govern themselves the same way that they are governing themselves—or at least to think that they are the safe ones among the reckless. In the end, however, this is a buyer-beware environment, where once they hit the road, each driving subject is accountable not only for their own safe driving but also for paying attention to and defensively protecting themselves from all the other driving subjects on the road as well. That's what makes driving the epitome of neoliberal subjectivity, but also why the figure of the car crash haunts the culture.

For the contemporary U.S. driving subject moving through this kind of physical and discursive landscape, therefore, to practice freedom and auton-omy within automobility means internalizing not only the value of autonomy and freedom but also the value of safety. Indeed, Jeremy Packer argues that the discourse of motorized mobility in the United States ultimately coheres around the "idea of safety," which "serves then as the solution" to resolving contradictions within the discourse at the same time that it "provides a nor-mative orientation for mobility." The discourse works through "techniques of docility and training" to produce a "'safe subject': a self that consciously and rationally reflects upon various practices and discourses regarding safety in the construction of its ethical subjectivity."[66] Because automobility grants freedoms and responsibilities to all drivers who pass minimal licensing procedures but does not directly control the behavior of these same drivers, the apparatus must produce safe subjects who value safety in their own driving and depend on other drivers also consciously or unconsciously working to make themselves over into "safe subjects" as well.

The main reason the system constantly produces accidents is that driving is a very ordinary activity that also demands extensive and ongoing vigilance: particularly vigilance about other drivers in the form of "lateral surveillance," but also vigilant "self-monitoring."[67] Jörg Beckmann writes that for automobiles to be safe on the road, they must be immobile in relation to one another even as they hurtle forward in relation to the road, where for each vehicle, "notwithstanding its movement through Euclidean (road) space, its position in network space is stable."[68] But this is exactly why "accidents happen" on the road and why car crashes are a symptomatic figure of social conflict haunting contemporary neoliberal societies: the network of drivers is a space of mobile privatization where each driver drives forward into the future with divergent goals and differential relationships to self-governing behaviors.

Especially in an era of road rage and moral panics about "impaired driving," this system is constantly on the verge of breaking down. Road rage is structured entirely as a frustration with other drivers who impede what is expected to be frictionless autonomous mobility. The same is true of complaints about traffic, where each person in the crowd looks to the others in the crowd as the problem instead of recognizing that the collective we call traffic is just an aggregate of individuals complaining about traffic. Likewise, the metaphorical and moral language of driver impairment assumes a normal condition of unimpairment. The system presumes that the natural state is where individuals are neutral in relation to other drivers and not driving while intoxicated (DWI) or "under the influence" (DUI) of some internalized external agent can be identified and thus regulated (especially alcohol, drugs, etc., but unlike sleep-deprived, careless, or otherwise inattentive driving) or "distracted" by some external agent that can similarly be identified and thus regulated: phoning, texting, etc. The discourse of DWI/DUI is perhaps the most direct encoding of neoliberal models of discursively produced individual accountability into the practice of driving: the price of admission into this system is the recognition that the system itself will never be to blame itself for the crashes produced within it.

As these examples demonstrate, the distinctive structure of automobility itself has contributed to its invisibility as a cultural discourse, particularly the way the act of driving feels empowering even while it is thoroughly embedded in institutional structures and collective frictions. And this structure has also kept people from recognizing that fatal car crashes are an intrinsic part of the system. As Sudhir Rajan puts it, "The car's very banality and the predictability of its institutions somehow serve as 'training wheels' to condition most of us to accept its importance without question."[69] I would argue that these same "training wheels" have conditioned drivers to structurally forget the risks of road trauma, which is part of what is combated by every road trauma shrine: each shrine refuses to allow the official cleansing of the crash site to erase the material reality of the crash, replacing the recognition of risk and costs into the same space that emphasizes mobility and freedom.

While automobility shapes many cultures around the world, it is particularly important in the United States, where driving a car has come to have central importance in materializing American ideals of democratic citizenship. As Jeremy Packer puts it in *Mobility without Mayhem* (2008), in the United States, "Being a good citizen means to be a self-empowered mobile citizen."[70] Likewise, Sudhir Rajan argues that in contemporary Western societies like the United States that produce "driver-citizens," automobility is intertwined with liberalism into an "automobility-liberalism combine" that mutually strengthens each apparatus and the interconnections between them. Rajan argues, "Automobility has become the (literally) concrete articulation of liberal society's promise to its citizens that they could freely exercise certain everyday choices: where they wanted to live and toil, when they wished to travel and how far they wanted to go."[71]

Indeed, Cotten Seiler points out in *Republic of Drivers* (2008) that from the earliest days of cars being mass-market commodities, the car's utility value and symbolic value were less important than the act of driving itself: "driving's sensations of agency, self-determination, entitlement, privacy, sovereignty, transgression, and speed; these were instrumental in establishing automobility as a public good and thereby ensuring its growth as an apparatus." Seiler's crucial point is that as the twentieth century developed more and more complex social, political, economic, and cultural systems that seemed to collude against any person's feeling of control or even comprehension, "driving still *felt* and *looked* like freedom"—and not only in idealized visions of automobility in car ads and road movies but also in everyday experiences of ordinary people behind the wheel.[72]

This points us to one final reason why road trauma has not registered as a collective trauma: as a form of sociality, automobility is extremely neutral in terms of social hierarchies, particularly in the United States. There are certainly class differences in terms of whether people are in private vehicles or mass transit, and whether they are pedestrians or bicyclists instead of drivers, when crashes occur. There are also class differences that translate into different likelihoods of drivers and passengers dying in car crashes in the first place, from the kinds of cars people drive, which also translates into access to safety features and the ability to maintain vehicles, to differences in access to quality health care if people do crash. And there are also class differences in access to less congested toll roads, which generally have fewer cars and fewer accidents. But beyond that, the streets and highways are one of the few places where people from every level of the larger society occupy the same physical and cultural spaces at the same time and do so with essentially equal power relations. The motoring public is an exceptionally diverse collective. Crashes can happen to anyone, anywhere, at any time. But because anyone could die and all kinds of people do die in road trauma, there are not built-in identity politics that people can draw on to see themselves as part of a preexisting collective identity, something that is almost always a prerequisite for social movements. We're all in it together, but by ourselves.

To appreciate the importance of automobility and the dynamics it produces, it is useful to relate it to the form of mobility-based citizenship automobility largely replaced in the United States: what we might call *locomobility* or mass mobility between predetermined locations.[73] Locomobility emphasized moving large numbers of people between defined locations on a precisely controlled schedule, first via steam train, stagecoach, omnibus, and ship and later via electric and diesel powered trains, subways, ships, and airplanes. It included not only huge corporate or governmental organizations that managed these systems but also encoded rigid class distinctions within the kinds of vehicles, destinations, and routes people had access to as well as the accommodations within those vehicles, destinations, and routes. Locomobility was dominant for almost a full century before automobility became dominant post–World War II, and remains today in the residual form of mass transit in large cities and commercial airlines between cities.

Where the automotive subject and automotive spaces mirror postmodern and postindustrial social and cultural values of decentralization, fragmentation, bricolage, identity politics, control, and neoliberalism, the locomotive subject and locomotive spaces mirrored modern values of centralization, uniformity, surveillance, and state control in the era of industrialization and mass culture. The locomotive passenger, whether they were in first class or below, recognized that they were subjected to the direction of hidden, expert drivers who worked for organizations that controlled every aspect of the infrastructure and structure of the experience. Passengers conformed their lives to schedules and routes determined and controlled elsewhere, so locomobility's relationship between autonomy and mobility produced a very different phenomenological reality than a private-vehicle, automobile-based system. That phenomenological reality of self-determined automobility became more and more focused on autonomy as the twentieth century progressed, especially as Fordist models of mass consumption evolved into post-Fordist models of mass customization and highway systems became more and more elaborate. The result was that as the types of cars proliferated and the infrastructure of automobility grew, drivers were given increasing freedom to choose not only where and when to go but also which kind of car to drive and what scale of road to drive upon.

On the other end of the timeline, contemporary emergent forms of spatial, economic, and cultural organization in relation to cars may soon either replace or significantly revise the now dominant form of automobility. You can see evidence of this shift in the development of sharing economy apps like Uber and Lyft and ZipCars, and in the current emphasis on developing viable electric and driverless cars. It's worth noting that these new technologies do not fundamentally challenge the implicit value for automobility but address some of its economic and environmental costs (the redundancy of car owner-ship, the reliance on fossil fuels, and the car's contribution to climate change) while ignoring others, such as the presumption of individual autonomy tied to mobility, land-use policy and the predominance of highway infrastructure to the American landscape, and the predominance of the car-based economy along highways and interstates. And finally, all of these trends are concen-trated among only certain demographics in certain urban areas. If you live in a city with access to ride-sharing and an infrastructure for charging electric cars, it is easy to ignore that the United States as a whole is still very much tied to the single-owner, single-driver, gas-powered model of automobility.

These trends suggest that if the United States has been a "republic of driv-ers," it might be moving to a new form of automobile subject tied to being a passenger as much as a driver, or at least a new kind of driver–passenger hybrid, where the "driver" drives decisions about where to go and when but a surrogate driver (machine or human) performs the mechanical task of driving the car itself. Like Uber and Lyft, computer-aided automobility suggests that the criterial element of automobility is *autonomous mobility*, not *driving* per se. Finally, competing forms of autonomous mobility not tied to automobiles, particularly velomobility, challenge the dominance of car-based automobility at one level but reinforce it at another: all contemporary systems of autono-mous mobility produce dispersed "drivers" who decide where to go and when to go, which means to use (an owned or shared car, bike, scooter, etc.), and which route to take in an even more dispersed system of governmentality. Looking it this way suggests that what had seemed the primary importance of owning a car in automobility in the twentieth century may have masked that the critical feature of automobility is *customizable self-determined car-based mobility*, not necessarily *who owns the car* or *how it is powered*.

And the crucial point I want to make here is that this naturalized associa-tion between individual self-propelled and self-determined automobility is exactly why car crashes are traumatic. Accidents happened within locomobil-ity, and people were traumatized by these accidents in a number of ways, but since train, bus, and later airplane crashes are never the fault of passengers, they come with a different phenomenological relationship to fate, and thus to trauma. Sometimes drivers and pilots are held accountable to protect the transportation company, but such accidents distribute cause to the system of locomobility itself more effectively than any accident within automobility does. Only a suicidal hijacker would get on a plane knowing both that they will die in an air crash and that they are to blame for an air crash. Every driver knows that people die in car crashes, but everyone keeps on driving, assuming it won't happen to them and especially won't be caused by them. Likewise, when crashes do occur, they come as a sharp surprise: a challenge not only to the integrity of one's body but also to one's identity as a mobile, autonomous driving subject.

Which brings us directly back to road trauma shrines. A car crash reveals both the breakdown *and* the apotheosis of automotive sociality—the collision of simultaneous copresence that demands an exchange, a crash of previously dispersed individual realities colliding into one another, a breakdown of one system and a breakthrough of another. And a shrine speaks of, to, and from this dynamic—significantly registering both a betrayal by and a belief in automobility. Their location on the roadside presupposes that people in automobiles will see them in roadspace, which itself presupposes continuing automobility. Car crash victims are witnessed in other settings, but there, on the side of the road, they are witnessed by other drivers driving by, doing the same thing that the victims were doing when they died: performing *and* being subjected to the creativities and constraints afforded by automobility.

To drive is to experience the freedom of autonomy and mobility—to be suspended in motion, to perpetually *be moving* between past and present, now and then, here and there—and making measurable progress toward a projected future. But to drive by a road trauma shrine is to be confronted with evidence of a different world hidden within the world of automobility, and to be reminded that mobility and autonomy are always structurally connected to the opposing forces of immobility and constraint. These forces are manifested in the largely disavowed prevalence of past, present, and future deaths and mayhem embedded in car culture.

Thus, as I will show throughout the book, the physical and discursive location of roadside shrines within spaces of automobility is a fundamental fact of their claim to a public. Indeed, in this, road trauma shrines further mirror the structure of automobility. When shrines are witnessed, they are witnessed not simply in the abstract but by other drivers—fellow members of the "motoring public." Importantly, these members of the motoring public are *doing* automobility on the very same road that already has claimed the victims memorialized in the shrines, and who thus, through *practicing automobility*, share an implicit affiliation, a collective identity, with the victim as drivers and/or passengers. If roadside shrines were located in a different physical and discursive space, they might make a different claim, and they might have a different relationship to collective trauma, memory, and witnessing, but the fact that they are located on the roadside ensures that their authority and their appeal always speaks to and from the discourses and subjects as well as the spaces of automobility.

Perhaps it is fitting that the car crash, the spectacular end to a life lived within a culture that presumes and even prescribes radically individual automobility, is figured individually in isolated roadside shrines as well. But the result is that as a form of collective memory, roadside shrines are so dispersed and apparently idiosyncratic that they are at first difficult to perceive as the *locations* of collective rememory, much less as *agents* of collective rememory. As spatially anchored stories of loss and defiance that transform a personal trauma into a public trauma, roadside shrines to car accident victims give visual, material, and spatial form to private memories that might otherwise be lost, ignored, or invisible. But they do more than perform rememory. In materializing rememory in the spaces strangers inhabit and traverse, they make certain demands on those who drive by them.

Roadside car crash shrines remind us that the trauma occurred, and they do so in a form structured like trauma and also in a material context contiguous with the original context of the trauma. That is, roadside shrines inscribe the automotive landscape with affective, traumatic memory sites where other people's memories intrude on our lives as drivers of public space unexpectedly, without us asking for them to do so. And they come in an evocative form, a visual/material/spatial form that shares with the language of dreams and trauma an iconicity that communicates visually and materially in excess of their explicit representational messages.

But, as we will see throughout the book, shrines do more than simply *insert themselves into* the public space of the roadside; once there, they *interpellate* a public as well. Present in other societies and cultures, they may materialize a different "message," but in the contemporary United States, they perform a silent and implicit challenge to the systematic forgetting of car crash deaths as well as the cultural dream of frictionless physical and social mobility that has fueled American culture in general since its infancy and American car culture in particular now for over a century. As they do, they bring the politics of collective affect into focus within the discourse of automobility, challenging drivers to remember that the risks of automobility are inscribed into the apparatus itself.

NOTES

1   Toni Morrison, *Beloved* (New York: Penguin/Plume, 1988).
2   Ibid., 35–36.

3 Ibid., 36.

4 For another example of how Morrison's rememory concept can be used to understand affective relations to memory and material culture, see Divya Tolia-Kelly, "Locating Processes of Identification: Studying the Precipitates of Re-Memory in the British Asian Home," *Transactions of the Institute of British Geographers* 29 (2004): 316.

5 Avery Gordon, *Ghostly Matters: Haunting and the Sociological Imagination*, 2nd ed. (Minneapolis: University of Minnesota Press, 2008), 201.

6 Ibid., 16; emphasis in original.

7 Jack Santino, "Performative Commemoratives: Spontaneous Shrines and the Public Memorialization of Death," in *Spontaneous Shrines and the Public Memorialization of Death*, ed. Jack Santino (New York: Palgrave, 2006), 5–16. See also Corrie Boudreaux, "Public Memorialization and the Grievability of Victims in Cuidad Juárez," *Social Research* 83, no. 2 (2016): 391–417; Margaret Burchianti, "Building Bridges of Memory: The Mothers of the Plaza de Mayo and the Cultural Politics of Maternal Memories," *History and Anthropology* 15, no. 2 (2004): 133–50; Rohit Chopra, "The 26/11 Network Archive: Public Memory, History, and the Global in an Age of Terror," *International Journal of Communication* 9 (2015): 1140–62; Erika Doss, "Death, Art, and Memory in the Public Sphere: The Visual and Material Culture of Grief in Contemporary America," *Mortality* 7, no. 1 (2002): 63–82; Erika Doss, "Spontaneous Memorials and Contemporary Modes of Mourning in America," *Material Religion* 2, no. 3 (2006): 294–319; Erika Doss, *The Emotional Life of Public Memorials: Towards a Theory of Temporary Memorials* (Amsterdam: Amsterdam University Press, 2008); Erika Doss, *Memorial Mania: Public Feeling in America* (Chicago, IL: University of Chicago Press, 2010); Jeffrey L. Durbin, "Expressions of Mass Grief and Mourning: The Material Culture of Makeshift Memorials," *Material Culture* 35 (2003): 22–47; Sylvia Grider, "Spontaneous Shrines: A Modern Response to Tragedy and Disaster," *New Directions in Folklore* 5 (2001): 1–10; Sylvia Grider, "Public Grief and the Politics of Memorial: Contesting the Memory of 'The Shooters' at Columbine High School," *Anthropology Today* 23, no. 3 (2007): 3–7; Peter Jan Margry and Cristina Sánchez-Carretero, "Memorializing Traumatic Death," *Anthropology Today* 23, no. 3 (2007): 1–2; Peter Jan Margry and Cristina Sánchez-Carretero, eds., *Grassroots Memorials: The Politics of Memorializing Traumatic Death* (New York: Berghahn Books, 2011); Ana Milosevic, "Remembering the Present: Dealing with the Memories of Terrorism in Europe," *Journal of Terrorism Research* 8, no. 2 (2017): 44–61; Ana Milosevic, "Historicizing the Present: Brussels Attacks and the Heritagization of Spontaneous Memorials," *International Journal for Heritage Studies* 24, no. 1 (2018): 53–65; Carmen Ortiz, "Pictures That Save, Pictures That Soothe: Photographs at the Grassroots Memorials to the Victims of March 11, 2004 Madrid Bombings," *Visual Anthropology Review* 2, no. 1 (2013): 57–71; Jack Santino, "Performative Commemoratives: Spontaneous Shrines, Emergent Ritual, and the Field of Folklore," *Journal of American Folklore* 117 (2003): 363–72; Jack Santino, ed., *Spontaneous Shrines and the Public Memorialization of Death* (New York: Palgrave, 2006); Gabrielle Soto, "Spontaneous Materiality: An Informal Survey of the January 2, 2011 Shrines at University Medical Center," *Arizona Anthropologist* 26 (2016): 76–98; and Marita Sturken, *Tourists of History: Memory, Kitsch, and Consumerism from Oklahoma City to Ground Zero* (Durham, NC: Duke University Press, 2007).

8 Santino, "Performative Commemoratives," 10, 13, 5.

9 Elizabeth Hallam and Jenny Hockey, *Death, Memory, and Material Culture* (Oxford: Berg, 2001), 181.

10 See especially Dennis Klass, Phyllis R. Silverman, and Steven L. Nickman, eds., *Continuing Bonds: New Understandings of Grief* (London: Taylor & Francis, 1996); Dennis Klass and Edith Maria Steffen, eds., *Continuing Bonds in Bereavement: New Directions in Research and Practice* (London: Routledge, 2017); Avril Maddrell, "Living with the Deceased: Absence, Presence and Absence-Presence," *Cultural Geographies* 20, no. 4 (2012): 501–22; and Jocelyn De Groot, "Maintaining Relational Continuity with the Deceased on Facebook," *Omega* 65, no. 3 (2012): 195–212.

11 Doss, *Emotional Life of Public Memorials*, 27.

12 See Cas Wouters, "The Quest for New Rituals in Dying and Mourning: Changes in the We-I Balance," *Body & Society* 8, no. 1 (2002): 1–27. See also Doss, *Emotional Life of Public Memorials*, 20; Doss, *Memorial Mania*, 80–93.

13 Candi Cann, *Virtual Afterlives: Grieving the Dead in the Twenty-First Century* (Lexington: University of Kentucky Press, 2014), 13.

14 Cann, *Virtual Afterlives*, 12.

15 Doss, *Memorial Mania*, 64.

16 C. Allen Haney, Christina Leimer, and Juliann Lowery, "Spontaneous Memorialization: Violent Death and Emerging Mourning Ritual," *Omega: The Journal of Death and Dying* 35, no. 2 (1997): 161.

17 Haney, Leimer, and Lowry, "Spontaneous Memorialization," 160. See Margaret Godel, "Images of Stillbirth: Memory, Mourning, and Memorial," *Visual Studies* 22, no. 3 (2007): 253–69; and Linda L. Layne, *Motherhood Lost: The Cultural Construction of Pregnancy Loss in the United States* (New York: Routledge, 2002).

18 Sturken, *Tourists of History*.

19 Wells, "Melancholic Memorialisation," 164.

20 Avril Maddrell and James D. Sidaway, "Introduction: Bringing a Spatial Lens to Death, Dying, Mourning, and Remembrance," in *Deathscapes: Spaces for Death, Dying, Mourning, and Remembrance*, ed., Avril Maddrell and James D. Sidaway (Farnham: Ashgate, 2010), 2 (emphasis in original).

21 Elizabeth Hallam and Jenny Hockey, *Death, Memory & Material Culture* (Oxford: Berg, 2001), 125–26.

22 Leonie Kellaher and Ken Worpole, "Bringing the Dead Back Home: Urban Public Spaces as Sites for New Patterns of Mourning and Memorialization," in *Deathscapes: Spaces for Death, Dying, Mourning, and Remembrance*, ed., Avril Maddrell and James D. Sidaway (Lanham: Ashgate, 2010), 169.

23 Ibid., 175. This is one of the main implicit political claims of spontaneous shrines. See Peter Jan Margry and Cristina Sanchez-Carretero, eds., *Grassroots Memorials: The Politics of Memorializing Traumatic Death* (New York: Berghahn Books, 2011).

24 See Jocelyn M. DeGroot, "Maintaining Relational Continuity with the Deceased on Facebook," *Omega* 65, no. 3 (2012): 195–212; Molly Hales, "Animating Relations, Digitally Mediated Intimacies between the Living and the Dead," *Cultural Anthropology* 34, no. 2 (2019): 187–212; Michael C. Kearl, "The Proliferation of

Postselves in American Civic and Popular Cultures," *Mortality* 15, no. 1 (2010): 47–63; Mirjam Klaasens and Maarten J. Bijlsma, "New Places of Remembrance: Individual Web Memorials in the Netherlands," *Death Studies* 38, no. 5 (2014): 283–93; Avril Maddrell, "Online Memorials: The Virtual as the New Vernacular," *Bereavement Care* 31, no. 2 (2012): 46–54; Alice Marwick and Nicole B. Ellison, "'There Isn't Wifi in Heaven!': Negotiating Visibility on Facebook Memorial Pages," *Journal of Broadcasting & Electronic Media* 56, no. 3 (2012): 378–400; and Nathalie Paton and Julien Figeac, "Muddled Boundaries of Digital Shrines," *Popular Communication* 13, no. 4 (2015): 251–71. For comparative studies of American material and virtual bereavement practices, see Cann, *Virtual Afterlives*; Connor Graham, Michael Arnold, Tamara Kohn, and Martin Gibbs, "Gravesites and Websites: A Comparison of Memorialisation," *Visual Studies* 30, no. 1 (2015): 37–53; Kate Sweeney, *American Afterlife: Encounters in the Customs of Mourning* (Athens: University of Georgia Press, 2014); and Tony Walter, Rachid Hourizi, Wendy Moncour, and Stacey Pitsillides, "Does the Internet Change How We Die and Mourn?: Overview and Analysis," *Omega* 64, no. 4 (2011), 275–302.

25   Nancy K. Miller and Jason Tougaw, eds., *Extremities: Trauma, Testimony and Community* (Urbana: University of Illinois Press, 2002), 1; see also Mark Seltzer, "Wound Culture: Trauma in the Pathological Public Sphere," *October* 80 (Spring 1997): 3–26; Mark Seltzer, *Serial Killers: Death and Life in America's Wound Culture* (New York: Routledge, 1998); Roger Luckhurst, "Traumaculture," *New Formations* 50 (2003): 28–47; and Ann E. Kaplan, *Trauma Culture: The Politics of Terror and Loss in Media and Literature* (New Brunswick: Rutgers University Press), 2005.

26   Cathy Caruth, "Trauma and Experience: Introduction," in *Trauma: Explorations in Memory*, ed. Cathy Caruth (Baltimore, MD: Johns Hopkins University Press, 1995), 3–12; Cathy Caruth, "Recapturing the Past: Introduction," in *Trauma: Explorations in Memory*, ed. Cathy Caruth (Baltimore, MD: Johns Hopkins University Press, 1995), 151–57; Cathy Caruth, *Unclaimed Experience: Trauma, Narrative, and History* (Baltimore, MD: Johns Hopkins University Press, 1996); and Bessel A. Van Der Kolk and Onno Van Der Hart, "The Intrusive Past: The Flexibility of Memory and the Engraving of Trauma," in *Trauma: Explorations in Memory*, ed., Cathy Caruth (Baltimore, MD: Johns Hopkins University Press, 1995), 158–82. See also Judith Herman, *Trauma and Recovery: The Aftermath of Violence from Domestic Abuse to Political Terror*, 2nd ed. (New York: Basic Books, 1997). For an early intellectual history of trauma studies, see Ruth Leys, *Trauma: A Genealogy* (Chicago, IL: University of Chicago Press, 2000).

27   Caruth, "Trauma and Experience," 4.

28   Kaplan, *Trauma Culture*, 34.

29   Roger Luckhurst, "Traumaculture," 28.

30   Ulrich Baer, *Spectral Evidence: The Photography of Trauma* (Cambridge: MIT Press, 2005), 9, 10.

31   See especially Cathy Caruth, *Unclaimed Experience: Trauma, Narrative, and History* (Baltimore, MD: Johns Hopkins University Press, 1996); Dominick LaCapra, *Writing History, Writing Trauma* (Baltimore, MD: Johns Hopkins Press, 2001); and Jeffrey C. Alexander, Ron Eyerman, Bernhard Giesen, Neil J. Smelser, and Piotr

Sztompka, eds., *Cultural Trauma and Collective Identity* (Berkeley: University of California Press, 2004).

32 Alison Landsberg, *Prosthetic Memory: The Transformation of American Remembrance in the Age of Mass Culture* (New York: Columbia University Press, 2004).

33 See Kaplan, *Trauma Culture*; see also Susan Sontag, *Regarding the Pain of Others* (New York: Picador, 2003).

34 See Emily Keightley and Michael Pickering, "Painful Pasts," in *Research Methods for Memory Studies*, ed., Emily Keightley and Michael Pickering (Edinburgh: University of Edinburgh Press, 2013), 151–66; Wulf Kansteiner, "Genealogy of a Category Mistake: A Critical Intellectual History of the Cultural Trauma Metaphor," *Rethinking History* 8, no. 2 (2004): 193–221; Wulf Kansteiner and Harald Weilnböck, "Against the Concept of Cultural Trauma," in *Cultural Memory Studies: An Interdisciplinary and International Handbook*, ed. Astrid Erll and Ansgar Nünning (New York: Walter de Gruyter Press, 2008), 229–41; Luckhurst, "Traumaculture," 28–47; and Antonio Traverso and Mick Broderick, "Interrogating Trauma: Towards a Critical Trauma Studies," *Continuum: Journal of Media & Cultural Studies* 24, no. 1 (2010): 3–15.

35 Keightley and Pickering, "Painful Pasts," 155–56.

36 For more perspectives on the relationship between a more aggregated form of cultural memory and a more rendered form of cultural memory, see especially Jeffrey K. Olick, "Collective Memory: The Two Cultures," *Sociological Theory* 17, no. 3 (1999), 333–48; and Avishai Margalit, *The Ethics of Memory* (Cambridge: Harvard University Press, 2002). My distinction between collected and collective memory also echoes the two types of collective memory identified in the 1920s by Maurice Halbwachs: collective memory in the form of individual memories formed alongside others within a certain social framework and collective memory as the shared memory of the group as a whole. See Maurice Halbwachs, *On Collective Memory*, trans. and ed. Lewis A. Coser (Chicago, IL: University of Chicago Press, 1992).

37 Jeffrey C. Alexander, "Toward a Theory of Cultural Trauma," in *Cultural Trauma and Collective Identity*, ed. Jeffrey C. Alexander, Ron Eyerman, Bernhard Giesen, Neil Smelser, and Piotr Sztompka (Berkeley: University of California Press, 2004), 1–30, 10. See also Ron Eyerman, *Memory, Trauma, and Identity* (New York: Palgrave MacMillan, 2019).

38 Luckhurst, "Traumaculture," 38.

39 Edward T. Linenthal, *The Unfinished Bombing: Oklahoma City in American Memory* (New York: Oxford University Press, 2003), 111.

40 See Paul Virilio and Sylvère Lotringer, *The Accident of Art*, trans. Michael Taormina (New York: Semiotexte, 2005). See also Arnar Árnason, Sigurjón Baldur Hafsteinsson, and Tinna Grétarsdóttir, "Acceleration Nation: An Investigation into the Violence of Speed and the Uses of Accidents in Iceland," *Culture, Theory & Critique* 48, no. 2 (2007), 199–217; Bednar, "Denying Denial"; and Charles Perrow, *Normal Accidents: Living with High-Risk Technologies* (New York: Basic Books, 1984).

41 Ann Cvetkovich, *An Archive of Feelings: Trauma, Sexuality, and Lesbian Public Cultures* (Durham, NC: Duke University Press, 2003), 3–4, 7.

42 Kathleen Stewart, *Ordinary Affects* (Durham, NC: Duke University Press, 2007), 39; see also Lauren Berlant, *The Queen of America Goes to Washington City* (Durham, NC: Duke University Press, 1997); Lauren Berlant, ed., *Compassion: The Culture and Politics of an Emotion* (London: Routledge, 2004); Eve Kosofsky Sedgwick, *Touching Feeling: Affect, Pedagogy, Performativity* (Durham, NC: Duke University Press, 2003); Mark Seltzer, "Wound Culture: Trauma in the Pathological Public Sphere," *October* 80 (1997): 24; and Mark Seltzer, *Serial Killers: Death and Life in America's Wound Culture* (New York: Routledge, 1998).

43 Stewart, *Ordinary Affects*, 39.

44 Ibid., 19, 57, 27.

45 Jill Bennett, "The Aesthetics of Sense-Memory: Theorizing Trauma through the Visual Arts," in *Regimes of Memory*, ed. Susannah Radstone and Katherine Hodgkin (London: Routledge, 2003), 28, 32–33; emphasis in original. For a more extensive treatment of this dynamic, see also Jill Bennett, *Empathic Vision: Affect, Trauma, and Contemporary Art* (Palo Alto, CA: Stanford University Press, 2005).

46 Bennett, *Empathic Vision*, 1.

47 Ibid., 3.

48 Ibid., 7; emphasis in original.

49 Bennett, "Aesthetics of Sense-Memory," 37.

50 National Highway Traffic Safety Administration, "NHTSA Confirms Fatalities Increased in 2012," accessed April 9, 2018, https://www.nhtsa.gov/press-releases/nhtsa-data-confirms-traffic-fatalities-increased-2012; and National Highway Traffic Safety Administration, "2016 Fatal Motor Vehicle Crashes: An Overview," accessed April 9, 2018, https://www.nhtsa.gov/press-releases/usdot-releases-2016-fatal-traffic-crash-data. More significant is the fact that "there were back-to-back total motor vehicle fatality increases from 2014 to 2015 (8.4%) and from 2015 to 2016 (5.6%)." For more detailed longitudinal data on fatal crashes in the United States, see also National Highway Traffic Safety Administration, "Fatality Analysis Reporting System (FARS) Encyclopedia," accessed July 5, 2011, http://www-fars.nhtsa.gov/Main/index.aspx.

51 National Highway Traffic Safety Administration, "Quick Facts 2016," accessed April 9, 2018, https://www.nhtsa.gov/press-releases/usdot-releases-2016-fatal-traffic-crash-data.

52 For a more international perspective, see Richard Dahl, "Vehicular Manslaughter: The Global Epidemic of Traffic Deaths," *Environmental Health Perspectives* 112, no. 11 (2004): 628–31.

53 See Michael Warner, *Publics and Counterpublics* (New York: Zone Books, 2002), 177. See also Doss, *Memorial Mania*, 94–96.

54 See Mikita Brottman, ed., *Car Crash Culture* (New York: Palgrave, 2001); and Paul Newland, "Look Past the Violence: Automotive Destruction in American Movies," *European Journal of American Culture* 28, no. 1 (2009): 5–20. The apotheosis of the American fascination with the spectacle of the car crash is J. G. Ballard's novel *Crash* (New York: Picador, 1973).

55 See especially Cotten Seiler, *Republic of Drivers: A Cultural History of Automobility in America* (Chicago, IL: University of Chicago Press, 2008); and Jeremy Packer, *Mobility without Mayhem: Safety, Cars, and Citizenship* (Durham, NC: Duke University Press, 2008). See also Daniel Miller, ed., *Car Cultures* (Oxford: Berg/ Bloomsbury, 2001); Peter Wollen and Joe Kerr, eds., *Autopia: Cars and Culture* (London: Reaction, 2002); Katie Mills, *The Road Story and the Rebel: Moving through Film, Fiction, and Television* (Carbondale: Southern Illinois Press, 2006); Steffan Böhm, Campbell Jones, Chris Land, and Matthew Paterson, eds., *Against Automobility* (Malden, MA: Blackwell, 2006); Brian Ladd, *Autophobia: Love and Hate in the Automotive Age* (Chicago, IL: University of Chicago Press, 2008); and Charissa N. Terranova, *Automotive Prosthetic: Technological Mediation and the Car in Conceptual Art* (Austin: University of Texas Press, 2014).

56 Paul Gilroy, "Driving While Black," in *Car Cultures*, ed. Daniel Miller (Oxford, Berg, 2001), 82.

57 Mike Featherstone, "Automobilities: An Introduction," *Theory, Culture & Society* 21, no. 4–5 (2004): 2.

58 See Mimi Sheller and John Urry, "The City and the Car," *International Journal of Urban and Regional Research* 24 (2000): 737–57; Mimi Sheller and John Urry, eds., *Mobile Technologies of the City* (London: Routledge, 2006); John Urry, *Sociology beyond Societies: Mobilities for the Twenty-First Century* (London: Routledge, 2000); John Urry, *Mobilities* (London: Polity Press, 2007); see also Tim Cresswell, *On the Move: Mobility in the Modern Western World* (London: Routledge, 2006).

59 Seiler, *Republic of Drivers*, 104; italics in original.

60 Seiler, *Republic of Drivers*; Packer, *Mobility without Mayhem*. See also Michel Foucault, *Power/Knowledge: Selected Interviews and Other Writings*, ed. Colin Gordon (New York: Pantheon, 1980); Michel Foucault, "Governmentality," in *The Foucault Effect: Studies in Governmentality*, ed. Graham Burchell, Colin Gordon, and Peter Miller (Chicago, IL: University of Chicago Press, 1991), 87–104; Michel Foucault, "The Ethic of Care of the Self as a Practice of Freedom," in *The Final Foucault*, ed. James Bernauer and David Rasmussen (Cambridge: MIT Press, 1988), 1–20.

61 Jeremy Packer, "Disciplining Mobility: Governing and Safety," in *Foucault, Cultural Studies, and Governmentality*, ed. Jack Z. Bratich, Jeremy Packer, and Cameron McCarty (Albany, NY: SUNY Press, 2003), 140. See Foucault, "Governmentality"; Gilles Deleuze and Félix Guatarri, *A Thousand Plateaus: Capitalism and Schizophrenia*, trans. Brian Massumi (Minneapolis: University of Minnesota Press, 1987), 474–500.

62 Packer, "Disciplining Mobility," 140; see also Graham Burchell, "Liberal Government and Techniques of the Self," in *Foucault and Political Reason: Liberalism, Neo-liberalism, and Rationalities of Government*, ed. Andrew Barry, Thomas Osborne, and Nikolas Rose (Chicago, IL: University of Chicago Press, 1996), 19–36.

63 Packer, "Disciplining Mobility," 148–49.

64 See Joseph R. Gusfield, *The Culture of Public Problems: Drinking-Driving and the Symbolic Order* (Chicago, IL: University of Chicago Press, 1981).

65 Beate Elvebakk, "Vision Zero: Remaking Road Safety," *Mobilities* 2, no. 3 (2007): 428.

66 Packer, "Disciplining Mobility," 136, 155; see also Peter Merriman, "'Mirror, Signal, Manoeuvre': Assembling and Governing the Motorway Driver in Late 1950s Britain," in *Against Automobility*, ed. Steffan Böhm, Campbell Jones, Chris Land, and Matthew Paterson (Malden, MA: Blackwell, 2006), 75–92.

67 See Mike Michael, "The Invisible Car: The Cultural Purification of Road Rage," in *Car Cultures*, ed. Daniel Miller (Oxford: Berg, 2001), 59–80.

68 Jörg Beckmann, "Mobility and Safety," *Theory, Culture & Society* 21, no. 4–5 (2004): 92.

69 Sudhir Chella Rajan, "Automobility, Liberalism, and the Ethics of Driving," *Environmental Ethics* 29, no. 1 (2007): 89.

70 Packer, *Mobility without Mayhem*, 5.

71 Rajan, "Automobility, Liberalism, and the Ethics of Driving," 88.

72 Seiler, *Republic of Drivers*, 9, 41, 67; italics in original.

73 For a full analysis of locomobility, including its impact on individuals, ideas, and landscapes, see Wolfgang Schivelbusch, *The Railway Journey: The Industrialization of Time and Space in the 19th Century* (Berkeley: University of California Press, 1986).

Chapter 3

Making Places for Performing Road Trauma

The single most distinctive aspect of road trauma shrines is that they are built as near as possible to the actual location of a car crash. Most crash victims are also memorialized in different places—in a gravesite at a cemetery or in a mausoleum, in home shrines, or in virtual memorials, for instance—but the site of the crash takes on intense material and spatial significance for mourners.[1] That is the most direct answer to the question of what road trauma shrines are doing on the roadside. Sometimes the person dies in the crash itself, and sometimes they die at the hospital or en route to the hospital, but the site of the shrine is always located at the literal and figurative "point of impact." The site becomes the material location of the trauma that occurred there, not only to the people killed in the crash but also the ones who experience the traumatic wrenching of their loved ones from their everyday lives.

Throughout the book, we will explore how these sites are the sites of trauma not only in the sense that someone died a traumatic death here but also because they mark the threshold where the crash victims were transformed from people going about their everyday activities to people who are "no longer with us," which is its own sort of trauma for those of us left behind. Here in this chapter, we will focus on how that is reflected in the process of place-making at road trauma shrines. Before we look in detail at the ways road trauma shrines make place, we should pause here to understand the general process of place-making within the landscapes of automobility more fully.

ROAD TRAUMA SHRINES AS PLACES
OF PUBLIC MEMORY

Cultural geographers generally define place in relation to its binary mate: space.[2] In this pairing, space is open, empty, undifferentiated, and unmarked, while place is heavy space, palpably significant, with a sense of its unique identity in relation to other places and surrounding space. In short, place is identified and invested space; it is a space of knowing, identity, and affective attachment. A place is a space of belonging, which also usually implicitly makes it a space of exclusion as well. Of course, whether a place is a place is not a question to be answered in the abstract. It is a question thoroughly entangled with cultural politics: as the history of colonialism indicates, one group's place is another group's space, and vice versa, depending on the cultural perspective that provides the entryway into or vantage point onto a particular location. Place as a concept also presumes a certain relationship to time—where space can become place over time (and vice versa), and is experienced differently by different people and groups at different moments of time—so questions about the *spatiality* of place are also about the *temporality* of place.

All places *contain* and *anchor* their own histories as spaces of extended inhabitation and signification, and all are in some form of proprietary relationship with the collectives that identify themselves as belonging to those places as the places belong to them. Moreover, places contain memories not only for groups but also for individuals within groups and individuals in multiple overlapping groups as well. However, for a place to be experienced as a "memory place" that shows that it is not simply a "place where something happened" or a "place that has a memory," a place must somehow be made to materialize and communicate memory in some structured, performative, and recognizable way.

Whether a place *conveys* and *performs* the memory it contains depends not only on the cultural politics of the person inhabiting the place but also the material realities of the site itself. In a global situation where the planet has been mapped extensively, there is no terrestrial surface that is not marked by humans in some way, but memory is performative: just as any space potentially can be a memory place, not all are identified or experienced that way. Thus, we might also locate the distinction between "place" and "memory place" temporally as a parallel distinction between abstract time and remembered time: places are actively marked and performed spaces, and memories are actively marked and performed times.[3]

Memory works at multiple scales of individual and social life and is simultaneously a thing and a process, so making sense of memory is extraordinarily complex. Making sense of the interrelationship between individual and group memory demands an even more nimble approach. Memory is neither a container of things nor a thing itself. It is a transactive process that produces a performative, momentary coherence of subject, object, past, and present. Memory is not a thing that can be remembered—put back together—in any literal sense. Memories repeat themselves, but never the same way. To remember is always to put together again (to remember) what has always already been and always will be again put together, temporarily, contingently, partially, problematically, and perspectively before.

Which brings us to the most vexing question in the interdisciplinary field of memory studies today, the subject of vigorous debates among researchers and theorists: where and how are memories—particularly traumatic memories—*located*? This question drives inquiry at the most micro of scales to the most macro of scales. With memories internal to the bodies of individuals, approaches to the question of where and how memory is (and is not) stored and processed in the body continue to be framed within psychoanalytic theory but are increasingly dominated by the discourse of neuroscience. With memories diffused to the scale of the society, culture, nation, diaspora, and transnation, approaches to the question of where and how groups remember and forget are framed within the disciplines of history, literature, sociology, anthropology, politics, communication studies, and geography. Often the question of where and how memories are located concerns the interrelationship of individual and collective memories, where the concern is not only for how the two are related but also procedural and political questions about the processes by which individual memories become collective memories, whose individual memories and group memories become part of larger collective memories, and whose individual and group memories are disavowed and forgotten in collective memories.

Theorists in the interdisciplinary field of memory studies since Maurice Halbwachs have generated multiple taxonomies of group memory, each with its own set of definitions and commitments: collective memory, cultural memory, social memory, public memory, postmemory, and prosthetic memory.[4] All seek to describe and theorize intersubjective forms of memory located outside of any one individual, where social relations, cultural discourses, and material constraints intervene even more visibly than they do within individual memories. At stake is not only understanding how individual memories are different from group memories but also how they interrelate and how they function. As Barbara Misztal puts it in her overview of theories of social remembering, "While it is an individual who remembers, his or her memory exists, and is shaped by, their relation with, what has been shared with others"; moreover, collective memory is also "always memory of an intersubjective past, of a past lived in relation to other people" within a particular social and cultural context who also continue to remember differently, together.[5] How these memories are shared and how that process is materialized within the spaces where memory is performed is the question, and it is a question best asked not in the abstract but, as Susannah Radstone and Bill Schwarz also advocate, while "working close to the ground," where we can analyze "historically specific formations of remembering and forgetting" where we find them, situated within their historical, cultural, and material contexts.[6]

When we work "close to the ground," we find that the ground itself is contested. As Kenneth Foote demonstrates, the American landscape is "shadowed ground," repeatedly inscribed, erased, and reinscribed with acts of violence and tragedy, some of which are remembered extensively and many others that are forgotten—either through intentional suppression or by the slow erosion of neglect.[7] This is where the cultural politics of collective memory emerge most clearly: at stake is not only who will be recognized as part of the collective but also who will be remembered in collective memory when they are traumatically torn from its fabric.

The difference between a trauma publicly remembered or forgotten at a particular site is determined by the culture's dominant regime of "necropolitics." Extending Foucault's understanding of biopower, Achille Mbembe has developed the term *necropolitics* to describe the way decisions about who lives and dies are "inscribed in the order of power," where some bodies and not others are deemed legitimately human and where some subjects and not others are empowered with "the capacity to dictate who may live and who must die."[8] Mbembe argues that the power to define who can legitimately kill and be killed is the foundation of state sovereignty: "To exercise sovereignty is to exercise control over mortality and to define life as the deployment and manifestation of power."[9]

But while Mbembe uses the concept of necropolitics explicitly to theorize the discursive production of actual bodies that kill and die, particularly within colonial spaces and the war on terror, the same thing happens with the *memory* of actual bodies of people who kill and die. This is true not only in colonial settings and in the war on terror but everywhere throughout contemporary regimes of public memory and forgetting, where there is a clear economy of power in which some subjects are deemed legitimately memorable and some are not, and where some subjects are legitimately allowed to memorialize their losses in public landscapes and others are not. The necropolitics of memory are currently most visible in contemporary conflicts about public memorials "left over" from the past, such as conflicts over Confederate war memorials: as necropolitics change over time, not only are different bodies remembered but also residual bodies become problematic.

With road trauma shrines, this move from bodies to memories is more than simply a metaphor, however. As I argue throughout this book, privately produced crash shrines develop, live, and die according to the logics of trauma, affect, mourning, and memory, which are all radically uncontained and unique to particular situations. You wouldn't know it from driving by the many roadside shrines in the United States today, but in most states it is explicitly illegal for private citizens to build and maintain a roadside memorial in the public right-of-way.[10] In these cases, state agencies assert a sovereign right to regulate crash memorials as unauthorized intrusions on the public landscape that they control on behalf of citizens in the name of "public safety." Given that they are rarely built legitimately, roadside shrines can never presume to exist in perpetuity the way an institutionally supported public memorial can.

Mbembe writes that "sovereignty means the capacity to define who matters and who does not, who is *disposable* and who is not."[11] Applied to the regulation of memorial objects within territories explicitly controlled by the state, we can paraphrase Mbembe to say: to exercise necropolitical sovereignty over memory in public landscapes involves deploying and displaying the power to control those spaces to facilitate some acts of remembrance and to obstruct or deny others. In such a regime, some subjects will be remembered, and some will be disposed of. Likewise, some private subjects will be allowed to memorialize their loss and others will not, and some memorial practices not originated by the state but seen as congruent with state objectives will be tolerated and some will not.

Compared to earlier American regimes of necropolitics, which made room only for remembering elites, the range of bodies considered memorable today in the United States has expanded greatly. Paul Williams argues that we live in "a time when every act of remembrance or emotional gesture is held, at least in the American context, to be precious—or at the very least non-disposable."[12] Similarly, in *Memorial Mania* (2010), Erika Doss argues that the contemporary United States is in the throes of "memorial mania"—a kind of "manic" and "excessive" compulsion to memorialize ordinary life in "visibly public contexts" that reveals a culture with an anxious relationship to time, history, and memory. Individuals and groups have mobilized to use public memorials as a "vehicle of visibility and authority" as they bring their concerns into public space.[13] These memorials themselves are "typified by adamant assertions of citizen rights and persistent demands for representation and respect" expressed through individual and subgroup "stories of tragedy and trauma," and are grounded in the politics of representation in the United States today—"in heightened expectations of rights and representation among the nation's increasingly diverse publics."[14] Importantly, this growing individualization of public memory not only expands the range of the memories being materialized but also challenges previous claims for a unified national memory and identity.

Even here, though, these newer individual-oriented public memorials are the outcome of collective political action and are ultimately institutionally sanctioned. This is not true for trauma shrines for urban murders or for road trauma, which are by definition the result of an extralegal, extrainstitutional process. The outcome is a contemporary memorial culture where all of these different memorial forms exist simultaneously next to one another: the official national war memorials, the hard-fought social movement memorials, the trauma shrines marking large-scale events, the trauma shrines commemorating public figures, the trauma shrines commemorating controversial deaths, and trauma shrines to functionally anonymous traffic fatalities.[15]

There is a growing recognition of the importance of analyzing these diverse "places of public memory."[16] Places of public memory are "implacably material" and immobile themselves—rooted in a unique geographical, topographical, and architectural context. But they also are often the destination of mobile visitors, who experience them rhetorically—not as "texts" to be read or seen, but as something engaging the entire human sensorium. Such places "act directly on the body in ways that may reinforce or subvert their symbolic memory contents," which has implications for analyzing their cultural politics and their histories as sites as well as the ways that visitors experience them rhetorically and phenomenologically.[17]

Places of memory include not only places that are site-bound from the beginning, such as trauma shrines, battlefield monuments, atrocity memorial museums, historical markers, and house museums, but also sites that become places through institutional effort, such as the memorials, monuments, and museums centralized on the Mall in Washington, DC. Because it is a memorial built at the site of the event it memorializes, the former type of memory place has an intrinsic, material relationship to the site. The political conflicts over such memory places are about whether they should be located where they are and who should determine their form and their content. Such conflicts arise because the sites usually have some other social function that they also need to continue performing: as functioning buildings, residences, parks, roads, etc. The latter sites also stimulate questions about whether they should exist and who should determine their form and content, but because they *could* be located anywhere, their siting becomes both more arbitrary and more obviously political.[18]

Once established, though, all places of public memory naturalize those initial choices and become places that can be experienced *as* places; they also then have their own histories as sites, where, as Blair, Dickinson, and Ott argue, "they do not just *represent* the past. They *accrete* their own pasts" as places.[19] The most sustained work in this regard has been with the Vietnam Veterans Memorial, where visitors have developed their own, now practiced, way of claiming the site as theirs through shrine activity, and the Park Service has incorporated these practices into curating the site.[20] Once they do that, they become memory places in a different way—as unique destinations, with their own sense of aura. An example that bridges this distinction is the Lincoln Memorial in DC, which was arbitrarily placed originally but has become a unique site-bound memory place itself as a site embedded within the public memory of the Civil Rights Movement and other social movements since.[21] That history has transformed the Lincoln Memorial into a place of public memory in every sense: it is a place of memory established to memorialize Abraham Lincoln, but it also is the place groups go to evoke Lincoln's memory to make public demands of the nation, and its identity as a place of public memory is now inseparable from this role.

Within the interdisciplinary field of public memory studies, there clearly has been more attention paid to institutionalized public memory places like the Lincoln Memorial than public memory places produced entirely by private citizens such as trauma shrines.[22]

Trauma shrines are places of public memory, but they maintain a unique identity within them. Their closest relatives are site-specific memorial museums, battlefields, and historical markers or history trails. Memorial museums are devoted to collecting and displaying images and artifacts meant to commemorate and instruct visitors on some singular historical event at the site of the past event, such as September 11th or Wounded Knee.[23] Some memorial museums, such as Holocaust museums in the United States, are not tied to a specific site where the event(s) happened, but most memorial museums are located at the sites being commemorated. Historical markers mark sites of institutionally determined importance, such as presidential childhood homes; historical trails trace routes of institutionally determined importance, such as the Mormon Trail or the Lewis and Clark Trail. Many times these forms converge, where markers are moved from their original site to the roadside where they will have a larger audience, or where a less formal pilgrimage "circuit" is designated to connect otherwise isolated sites or memorial museums into an extended geography of remembrance for a large-scale event, such as the many sites devoted to commemorating the Civil Rights Movement in the South, Holocaust memory sites throughout Europe, and Japanese-American Internment Camps throughout the American West.

What distinguishes these related memorial forms from road trauma shrines is first and foremost their institutional context. These other memorial forms all originated and are maintained within a certain institutional apparatus and within particular bureaucratic structures and processes. Certainly, ordinary people advocated for building many of them, but they are usually designed by professional artists and are not built unless they are vetted through multiple institutional processes, which ensures that they have a central set of intended functions and an identified (and often clearly articulated) sense of purpose. And once they are built, they become public property owned by everyone and no one at the same time, which serves as the shared foundation for both celebrations of and resistance to their presence and form.

Conversely, by definition, trauma shrines are produced by private citizens who originate, design, and maintain the sites without coordinating with any kind of governmental agencies. Shrine builders thus also maintain a strong sense of ownership of the site, even though in most cases the shrine is not placed there legitimately. Unlike more official acts of rememory that officially honor public and private citizens, such as manifestly social site-defined national memorials at Ground Zero in New York City and in Oklahoma City, war memorial sites at Gettysburg and Pearl Harbor, and the ubiquitous state and national roadside historical markers, roadside car crash shrines are *vernacular* acts of rememory through and through. Not only are they produced by people not invested with institutional authority, they are produced "all over the place." They are not located in a singular place that archives and curates displays of accidents to the public through some institutional apparatus officially designated with the power to do so. Instead, they are radically dispersed and collectively authored.

The closest analogous forms of road trauma remembrance similarly institutionalized are seen in Mothers Against Drunk Driving (MADD) crosses and state highway markers that promote safe driving or designate responsibility for litter control (Adopt-a-Highway) while also memorializing a crash victim, and even these markers are never "collected" into a circuit or trail for potential trauma tourists the way historical markers often are. As we will discuss much more fully in chapter 6, institutionalized roadside memorial forms like these are almost always tied to generalized public safety concerns about impaired driving in addition to remembering certain private citizens, so they not only subsume the individual loss to an abstracted collective goal but also usually share with memorial museums a presumed discourse of perpetrator/victim. In contrast, privately built road trauma shrines almost always presume victimhood without naming a perpetrator, where the people being commemorated are characterized as the victims of circumstance and fate—the unfortunate and unexpected recipients of an accident rather than the outcome of the malign intent of a perpetrator.

Many memory places are manifestly public, in the sense that they are institutionally controlled by the state, which is charged with providing for the use of and protection of these spaces on behalf of a local, regional, national, or transnational public. Many are also public in the more general sense of accessible to all and collective and shared as well. Many are thus also the focus of contemporary trauma tourism or disaster tourism, the practice of visiting sites of local, national, or global trauma: pilgrimage sites for citizen-tourists to visit and experience as a rite of citizenship-based witnessing or political commitment.[24]

But unlike these institutionalized places of memory, which are destinations that draw strangers who seek them out and who do so alongside other visitors, road trauma shrines are often encountered by strangers *accidentally* as they drive by them, and rarely in groups of other strangers except early in their life at vigils and later at important anniversaries.[25] Friends and family gather at shrines, where they mourn in an obviously collective setting (including even those who may not know each other but who do know the person memorialized at a shrine), but this rarely happens among strangers at roadside shrines. Road trauma shrines place memory, each creating what Pierre Nora calls a *lieu de memoire*—a particular intentionally produced place where memory both "crystallizes and secretes itself" in material form through multiple visual, material, and spatial means.[26] However, unlike Nora's sites, crash shrine sites are not always already *national* sites. They remember citizens without a clearly collective identity in a space that by itself makes no apparent claim to collectivity as well. Where some violent deaths can be made meaningful as cultural sacrifices by using shrines to recuperate private deaths within discourses of nation and citizenship, the deaths of ordinary people who die in car crashes resist sensemaking at the national, global, or even local scale.

MAKING PLACES FOR PERFORMING ROAD TRAUMA

Now that we have a better sense of the kind of memory place a road trauma shrine is, we need to more specifically analyze the ways they make place. As Gregory Ulmer argues, "Traffic fatalities are fundamentally 'abject'": if they perform a cultural sacrifice on behalf of some larger cultural value, that value "remains inarticulate, within the bodies and behaviors of individuals in the private sphere, untransformed, nontranscendent, unredeemed."[27] In the absence of some larger public apparatus designed to shift each act of memory from "the sphere of one-at-a-time individual personal loss to the public sphere of collective identity," individual mourners remembering road deaths are left to take matters into their own hands.[28] But it is more than that. The very feature that gives a roadside shrine its material affect—its unique spatial relationship to a unique site of trauma—is the thing that keeps car crash memorial practices dispersed, and thus mitigates seeing them as a collective form. Moving the memorial elsewhere would negate the shrine's function as a spatially unique portal between the living and the dead, where mourning intervenes in and at the site of trauma.

On the other hand, as I will argue throughout the book, the dispersal of road trauma shrines might indeed be the most effective way of registering the cultural trauma they figure. Road trauma shrines are direct evidence of the ongoing ordinary trauma of living in a culture entangled with car-based automobility. They are an intrusive presence in the roadscape—an affective, transactive reminder of the everyday traumas of automobility. But since they intrude, they cannot be either contained or avoided. If they could be contained in a single place, such as something like a singular National Car Crash Memorial, they could just as easily be ignored as celebrated. But because they are out there on the roads, where strangers encounter them without asking to do so, they become affective reminders that we consciously or unconsciously either ignore or disavow the prevalence and immanence of automobile deaths as we continue to drive forward. There they sit, these silent witnesses to another reality that won't go away, no matter what we may wish, asking us to remember that we forget to see the car accident not as an aberration but as a systematic outcome of a culture that has placed its bets on the side of automobility.

Together, road trauma shrines comprise an embodied, material refusal to either ignore car crash deaths or accept car crash deaths as a matter of course, as collateral damage to a culture committed to cars. Together, they demand witnessing on the roadside, where drivers driving by them are forced to experience them as something intimately connected with their own mobility, if only for a moment. Together, surrounded alternately by the "open" spaces of the roadside or by the density of the sprawling everyday commercial landscape or even the density of "home," a road trauma shrine demands that we avert our eyes from the windshield and our tomorrow-bound lives for long enough to feel consequences—and the past—refusing to disappear from the rearview mirror.

Even the most apparently chaotic, spontaneous, temporary, and "makeshift" shrine is organized elaboratively internally and in relation to its immediate surroundings, and each also reflects a larger pattern of spatial practices—and a larger pattern of inhabiting the material and discursive spaces of automobility as well as trauma and memory. Sometimes shrines seem to fit in to their surroundings, and other times they seem more obviously dislocated. Regardless, they often display their difference from their immediate surroundings by marking out their territory to make this rupture materially present. Sometimes the shrine produces its place right next to or even incorporating obvious evidence of the crash itself.

But whatever form they take, each roadside shrine is always a *tactical* phenomenon, a temporary emplacement located inside a highly circumscribed and complexly shared public space. Here I am drawing on de Certeau's distinction in *The Practice of Everyday Life* (1984) between *strategy* and *tactics*.[29] For de Certeau, strategy is the tool of institutionalized authority, which is invested with the sovereign right to not only establish places but also determine which uses of a place are appropriate and legitimate. As de Certeau puts it, a "strategy assumes a place that can be circumscribed as *proper*."[30] On the other hand, the tactical use of space temporarily *appropriates* places, "poaching" the place of strategy to temporarily make space for a precarious new place. In this formulation, "The place of a tactic belongs to the other. A tactic insinuates itself into the other's place." As it does, "Whatever it wins, it does not keep."[31]

Each road trauma shrine is a tactical taking of a place that does not belong to the shrine builder: the public right-of-way (ROW), which belongs to the "public" but is produced as proper as it is controlled strategically by some institutional authority on the public's behalf. However, this tactical taking is predicated on a different claim of ownership: shrine builders do not strategically control the place, but the trauma that occurred there has given them a different authority to take the place as their own. This tactical character is important to keep in mind as you think about how road trauma shrines are situated.

Not every crash death is memorialized in a shrine, but when a shrine is established, multiple authors carve out space in the public ROW to claim and inscribe an otherwise unmarked (or at least differently marked) piece of ground with material signs of implicit trauma, grief, and memory focused on one or more individuals killed in a crash at the site. There are four features of place-making at road trauma shrines: they are usually either set up on the roadside itself (*Roadside Islands*) or attached to the existing roadside infrastructure (*Roadside Attachments*). A *Roadside Island* makes place by building a site from scratch on the roadside, whereas a *Roadside Attachment* makes place by attaching the site to an existing object or piece of highway infrastructure in the ROW, such as a guardrail, fence, road sign, tree, or utility pole. In addition to this, all road trauma shrines have distinctive patterns of *Orientation* to the road and drivers on it, and to *Evidencing the Crash* itself, in the form of damage to the site itself caused by or associated with the crash, such as skid marks, guardrail damage, tree scars, and police investigation markings.

Shrines generally make place in one of two ways: they are either *Roadside Islands*, built on the open roadside or the median, or *Roadside Attachments*, built onto some other piece of existing roadway infrastructure. Of the hundreds of road trauma shrines I have witnessed, only one of them took a different form than this. This site, in Southern California, has a shrine on the side of the road, built into the rocks just next to the road, but also inscribes the road itself with a remembrance of the victim: "R.I.P. DAD" (see figure 3.1). The spray-painted words on the road evoke the spray-painted police investigation markings at many crash sites, but here are used to remember the victim instead of map out the crash. The site also is distinctive in its scale: it includes not only more common elements—such a Virgin Mary statue, teddy bears,

Figure 3.1. *Roadside and roadway shrine*, California State Highway 94 East, Indian Springs, CA, USA, July 2006.

and flowers—but also numerous spray-painted messages on rocks on *both* sides of the road as well as *on* the road itself. Also distinctive is the direct address of the "Slow Down" message on the side opposite the main shrine. The effect is intriguing: drivers materially drive over, in between, through, *and* next to the shrine, ensuring a much more embodied encounter with the shrine than most, which are usually experienced by drivers as something visual framed through their windshield or one of their side windows.

ROADSIDE ISLANDS

Roadside Islands are shrines that are built directly on the ground, either on the side of the road or in the median between parts of a road. Some have very loose borders, but most have a clear border, which functions in two directions at once: the border establishes an inside from an outside, enclosing the shrine so that there is a clear boundary both keeping the shrine in and keeping the rest of the surrounding landscape out. The effect is that they look to me like little island cities in a sea of land, which makes them literal manifestation of Astrid Erll's observation that "memories are small islands in a sea of forgetting."[32]

This form of shrine is particularly prevalent in the desert Southwest, where foliage is already more sparse and close to the ground. In such an environment, an island shrine stands out rather starkly from its surroundings, especially making shrines located on long, straight roads visible for several miles. It is clear that builders of shrines often transport the main objects of the shrine, including even the gravel or mulch often used to suppress weeds and mark space within the border, but also use rocks found nearby to mark the boundary or to hold a cross upright in hard, dry ground. Sometimes, in addition to using indigenous materials to form the border of the shrine, people

Figure 3.2. Top Left: New Mexico State Highway 68-South, Embudo, NM, USA, August 2003. **Top Right:** Arizona State Highway 85-South, North of Why, AZ, USA, August 2006. **Bottom Left:** Arizona State Highway 85 @ Arizona State Highway 86, Why, AZ, USA, August 2006. **Bottom Right:** U.S. Highway 285-North, South of Española, NM, USA, February 2010.

place local stones in recognizable shapes, such as crosses, peace signs, hearts, or spirals; other times they use stones to spell out words, especially "R.I.P."

The American Southwest has a diverse range of microenvironments determined mostly by proximity to bodies of water or elevation above sea level. In more lush Southwestern environments on the Pacific Coast, in central and east Texas, or high in the Rocky Mountains, for instance, a shrine's border must be continually maintained through mowing the immediate surroundings of the shrine, or suppressing surrounding vegetation through chemical herbicides or more elaborate landscaping.

ROADSIDE ATTACHMENTS

The second main form of roadside shrine is what I call a *Roadside Attachment*. Instead of being built up from scratch on the ground, these shrines are attached to some preexisting part of the roadside infrastructure or surrounding landscape: guardrails, utility poles, road signs, traffic lights, commercial

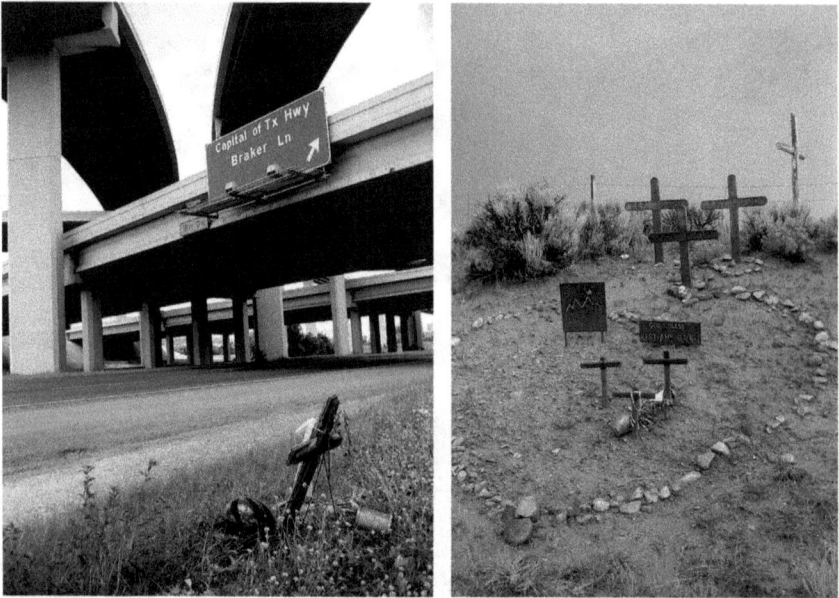

Figure 3.3. Left: Loop 1-North @ U.S. Highway 183, Austin, TX, USA, July 2003. Right: U.S. Highway 64-East, West of Taos, NM, USA, August 2003.

signs, fences, and trees. Because they are attached to existing objects, their relation to the roadside infrastructure is usually more visible than with Road-side Islands. In densely populated urban areas, where shrines interpellate both pedestrians and drivers, attachment shrines sometimes will circle entire utility poles and trees, but in most cases, attachment shrines use the existing infrastructure as a kind of platform to install a shrine that is oriented in one direction: toward the road, either facing the traffic closest to it or facing the road more frontally. We will look at patterns of *Orienting* shrines to the road later in a separate section. In some cases, the shrine is attached to the infra-structure object that the vehicle crashed into. In this case, the object provides not only a platform for the shrine but also *Evidence of the Crash* itself, which gives the siting of the shrine an added material dimension that we will discuss more fully later in a separate section as well.

ORIENTATIONS

Whether it is a Roadside Island or a Roadside Attachment, almost every road trauma shrine is spatially polarized with a clear front and a clear back that is oriented so that it faces drivers. This might seem like an obvious observation,

Figure 3.4. Soledad Canyon Road-East, Overlooking California State Highway 14 (Antelope Valley Freeway), Humphreys, CA, USA, July 2006.

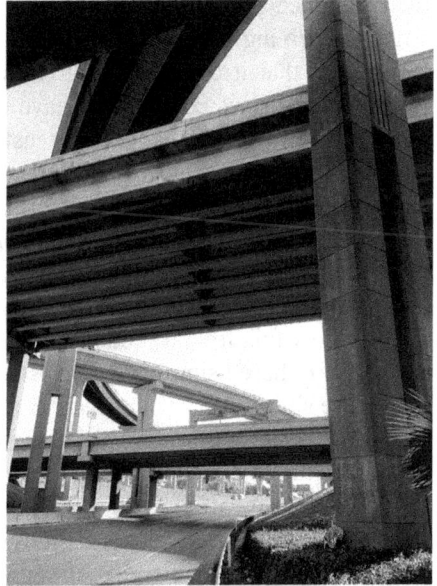

Figure 3.5. Left: Interstate 70-East, West of Green River, UT, USA, July 2006. Right: U.S. Highway 290 East at Interstate 35 South, Austin, TX, USA, March 2017.

Figure 3.6. New Mexico State Highway 76-South, West of Cordova, NM, USA, August 2003.

but that is only because it seems to be natural given what we know about the dual audience of road trauma shrines—that they address both intimates and strangers driving by. As with any form of cultural analysis, it is important to imagine that it could be different so we can determine what is actually significant. Shrines could either have no clear pattern of orientation or they could face away from the road, for instance. The fact that shrines all face the road in some way is indeed the clearest material evidence that road trauma shrines have an ambition to reach larger publics than the people who already knew of the loss. That is, instead of only serving as places for people who already know about the trauma and the shrine it helps negotiate, shrines are always set up so that they address strangers driving by them as well. We will analyze how this frontal mode of address interpellates multiple publics later in chapter 6; here I simply want to establish this as a distinctive feature of the way these shrines carve out space to make place in the ROW.

Indeed, because they are built to be seen by drivers, shrines face the traffic the same way that everything and everyone on the road addressed to drivers faces the traffic: official signage (stop signs, speed limit signs, etc.), billboards, real estate signs, lost pet signs, garage sale signs, panhandlers, hitchhikers. Like these other forms of addressing drivers, shrines are speech acts that call out to people not necessarily looking for them, and make something happen when they ask drivers to pay attention to them.[33]

Incidentally, that is actually also the main reason cited when state agencies regulate or remove privately built roadside shrines: that they are a distraction to drivers that create a public safety hazard. Shrine builders rightly reject this rationale as specious, noticing the same thing I mentioned above about other modes of roadside address and asking why shrines are regulated or banned when other forms of distracting drivers—ironically even including official messages warning drivers about crashes—are promoted. Here it is important to see that they would not be seen as a public safety hazard if they were not presumed to be distracting to drivers, and they wouldn't be presumed to be distracting if they were not seen to be successful in asking for drivers' attention and communicating with them along the road.

In the hundreds of road trauma shrines I have seen in the United States and beyond, I have seen only a few exceptions to the pattern of shrines facing traffic.[34] The main exception is when a shrine is placed in the median of a divided highway or street, where shrines often have two faces—with the shrine set up perpendicular to the roadway with one side facing each line of traffic going in opposite directions. Similarly, shrine builders working with a controlled-access highway with a frontage road orient the shrine either toward the main traffic lanes or to the frontage road. The only time this difference matters is when a shrine is parallel to the line of traffic. When the shrine orientation is perpendicular, the space between the main traffic lanes and the frontage road functions as a second median, and the shrine faces both lines of traffic going in the same direction.

Figure 3.7. U.S. Highway 285-North, South of Roswell, NM, USA, August 2003.

I have only seen one shrine on the side of the road that faces two directions at once (see figure 3.7). This shrine, in New Mexico, memorializes two people killed in the same crash, one for each side of a cross that was installed perpendicular to the road on a two-lane highway. In this example, based on what I have already observed about the general pattern, you could still make the argument that the Ortiz side of the double-sided cross is implicitly made more salient by the fact that it faces the oncoming traffic.

Exceptions like this are rare, however. Most road trauma shrines are built on the side of the road instead of a median, and shrines on the side of the road always either face the road directly (parallel to the road) or face the line of traffic on the same side of the road (perpendicular to the road). This is true for long-established shrines, but it is also true even when the shrine is actively being built and not yet fully formed into a separate place. At most other shrine sites in the United States, especially in the American Southwest, the clearest indicator of orientation is actually the cross, which is often made into the central element, and is also the place where a photographic portrait of the victim or some other proxy representation of the person, such as a MOM or DAD floral arrangement, is affixed. We will discuss the role of the cross as a symbolic object in its own right in the next chapter, but for now it is important to establish that when there is a central cross, the cross always has a clear front, and it always faces the road.

EVIDENCING THE CRASH

Another distinctive feature of road trauma shrines—apparent in both the Island form and the Attachment form—is that they either explicitly or implicitly establish a relationship between the shrine and other material evidence of the crash. We will analyze the debris itself more extensively in chapter 4. There, I will show how specific pieces of crash debris function as relics, serving to materially connect the victim's body and car as prosthetics to the shrine. Here I want to focus on how these relics and other material evidence of the crash can become powerful forces in the process of place-making.

Jörg Beckmann argues that "automobility 'works,' *because its accidents are denied.* Collective denial enables individual mobility."[35] Beckmann reasons that "if it wasn't the subject of denial, the wreck would simply be left in the ditch, as a testament to the dangerous aspects of driving along that particular stretch of road"—as a kind of *memento mori* displaying the risks of driving to discipline its drivers.[36] In lieu of a *Mad Max* landscape strewn with the evidence of millions of car crashes, we have car crash shrines. Indeed, Mike Featherstone writes that roadside shrines speak back to the official control of the site that removes the crash to make way for new mobilities, where expert discourses "continually run up against the public's experience of the immediacy of automobile deaths and injuries and protest at the irrational sacrifice of life." Shrines thus materialize an implicit protest against systemic denial, where shrines "endeavor to inscribe the site as a place of tragedy and remembrance, by refusing to erase the incident from public memory and allow drivers to relax back into the normal traffic flow."[37] While the institutionalized process performs its own form of collective memory of road trauma and automobility, roadside shrines show the limits of such an approach and demonstrate instead that the culture needs to perform these memories differently.

Each time an accident scene is "cleansed" by police, EMS, and road crews, the evidence of the ordinary risks of driving is materially denied. In a system where drivers' rights to autonomous mobility are experienced not as conscious choices but as naturalized, taken-for-granted citizenship rights guaranteed not only by the license to drive but also the participation in the dominant culture, the naturalizing of automobility in general and driving as a performative practice is dependent on denying the risks to self and others implied in the system from the beginning. By rendering evidence of the lack of system safety invisible, the crash cleanup not only reasserts the discourse of "public safety" but also reinforces the discourse of automobility by amplifying the structured forgetting of the everyday traumas embedded in automobility.[38]

Practices of site cleansing vary widely across municipalities and states, where in some cases no evidence of the crash site remains after the vehicles have been removed. At most crash sites, however, there is at least some debris left over, especially early in the life of a shrine. There are usually material signs of the crash carved into the ground and surrounding roadscape, including the vehicle debris left over from the crash, visible scars to road infrastructure or vegetation, land damaged by the crash and the removal of the vehicle(s) involved off pavement, and markings made in the police investigation of the crash. Regardless of whether shrines acknowledge and even incorporate these things or exclude them from the shrine site altogether, they give material form to the violence that occurred there, certifying the trauma and speaking back to the denial of it that is central to automobility.

Of course, the official "cleansing" of deadly road crashes often includes police and insurance investigations that leave their own traces in the roadscape, especially when reclaiming the site involves investigators tracing the path of the accident. The result is ghostly spray-painted outlines of vehicles where they collided with other cars or roadside objects and came to rest, spray-painted skid paths, and sometimes even the resonant letters "POI," or Point of Impact, sprayed on the pavement.

Figure 3.8. Left: U.S. Highway 290 @ U.S. Highway 183, Austin, TX, USA, October 2015. Right: U.S. Highway 290-East, East of McDade, TX, USA, November 2012.

Figure 3.9. Top Left: Airport Boulevard @ MLK Boulevard, Austin, TX, USA, February 2010. Top Right: Texas State Highway 71-West, West of Bee Cave, Texas, USA, February 2008. Bottom Left: California State Highway 67-North @ Scripps Poway Parkway-East, South of Ramona, CA, July 2006. Bottom Right: Texas State Highway 71-West, West of Bee Cave, Texas, USA, February 2008.

Shrines sites often exist side by side with these inscriptions, making it evident that there are two kinds of inscriptions going on at these sites: the institutionalized police mapping of the crash and the shrine makers' mapping of their trauma to the material objects in the shrine. The official inscriptions perform their own kind of speech act: *here is where the crash happened and how.* They mark the space as significant in a different but no less material way than the shrine, and they have their own code, often delineating the trajectory of colliding vehicles as well as the final orientation of vehicles after a crash. Materially, they demonstrate the control of the ROW by the state, showing that someone other than the drivers involved has the authority not only to determine which facts are important enough to be entered into evidence but also to use spray paint to inscribe the space with impunity, where it would be considered graffiti if anyone else did it. Most important, where shrines focus on showing the person who was lost in the crash, these alternate inscriptions emphasize the violent *crash*, not the *person lost* in the crash, which makes for an intriguing tension when both inscriptions are present at particular sites.

That is what is happening in the following example from a crash site near San Diego, California, where the cross-shaped police markings for the wheel locations on the road have an uncanny mirror in the form of a cross on the roadside next to them (see figure 3.10, Top). Such an image provides an icon of the often conflicted interrelationship between the authority of grieving subjects and the authority of the state.

In some cases, the material traces of the crash itself and the subsequent police investigation outlive the shrines that once occupied the same space, making them the only visible way for strangers to know about the trauma that once occurred there. That is evident in a site just a few hundred yards north of the previous example near San Diego, where the police markings, charred roadside, and charred remains of the vehicles involved in the crash are even more vivid reminders of the trauma than the faded plywood and discarded planters left over from an abandoned shrine (see figure 3.10, Bottom).

One day these markings too will fade, but as long as they exist, their presence is a clear reminder not only that a car crash happened but also that it was significant enough to mobilize the state apparatus around building a theory of cause, effect, and blame. They remind us that an investigated car crash encodes a different kind of place-bound "public memory" than a roadside shrine: the official determination of personal responsibility or innocence represented first in the police report and later in court documents and rulings, which, along with parallel insurance company investigations, seek to render an officially sanctioned "public" interpretation of the crash—especially when there are criminal charges or civil damages at stake.

Of course, this is not the kind of public I have in mind when I argue that roadside shrines interpellate multiple publics and have the potential to form new collective memories of road trauma. Indeed, the process of officially determining or denying individual responsibility in a car crash is the most forceful form of reprivatizing of death that can happen in the case of a car crash: whether through absolution or conviction of individual drivers, the official determination of cause externalizes the accident from the system and

Figure 3.10. Top: *Police Markings on Roadway Next to Shrine*, California State Highway 67-North @ Scripps Poway Parkway-East, South of Ramona, CA, USA, July 2006. Bottom: *Charred Road and Police Outline of Left-Rear Wheel*, California State Highway 67-North @ Scripps Poway Parkway-East, South of Ramona, CA, USA, July 2006.

places it into the realm of individual actors, which always lets the cultural apparatus of automobility off the hook.

Road trauma shrines are not only a *feature* of automobility but also a defining *figure* of automobility. They are rooted in a physical geographical place and commemorate an event that occurred in a very definite space and time. The sites reinforce a larger sense of time: dwelling in place, being, remembering, not forgetting. But they also mark the spot where someone's desire for, expectations for, and practice of autonomous mobility very literally came to an end. In this regard, the highway itself is both a central cultural form and a central metaphor of the simultaneously local and global nature of contemporary culture, where the here/now and there/then always exist together. Highways are a distinctive kind of social space with a distinctive sociality. They are fully emplaced—they exist in particular places and serve particular local needs, etc.—but they also are places for mobility, places that people move through instead of dwell in, places that create locales as they connect to places far from the local.

Pull out of your private driveway, and you could be on the way to anywhere on the continent. This duality is a significant feature of the highway deaths memorialized by road trauma shrines: people on their way from somewhere to somewhere else had their journey to elsewhere cut short, redirected, and ended at a particular place where the road goes through but itself does not stop. And it is exactly at this peculiar sort of place that strangers driving by encounter them.

NOTES

1 For an analysis of how roadside shrines relate to their corresponding gravesites, see Anna Petersson and Carola Wingren, "Designing a Memorial Place: Continuing Care, Passage Landscapes and Future Memories," *Mortality* 16, no. 1 (2011): 54–69. See also Holly Everett, *Roadside Crosses in Contemporary Memorial Culture* (Denton: University of North Texas Press, 2002).

2 The classic enunciation of this distinction is Yi-Fu Tuan, *Space and Place: The Perspective of Experience* (Minneapolis: University of Minnesota Press, 1977). See also Tim Cresswell, *Place: A Short Introduction* (Malden, MA: Wiley/Blackwell, 2004); and Phil Hubbard and Rob Kitchin, eds., *Key Thinkers on Space and Place* (London: Sage, 2010). Two important contemporary counter-models are Henri Lefebvre's pairing of natural "absolute space" and social "representational space" and Deleuze and Guatarri's pairing of "smooth" and "striated space." See Henri Lefebvre, *The Production of Space*, trans. Donald Nicholson-Smith (Oxford: Blackwell, 1991); and Gilles Deleuze and Félix Guatarri, *A Thousand Plateaus: Capitalism and Schizophrenia*, trans. Brian Massumi (Minneapolis: University of Minnesota Press, 1987), 474–500.

3 Cf. Carole Blair, Greg Dickinson, and Brian L. Ott, "Introduction: Rhetoric/
Memory/Place," in *Places of Public Memory: The Rhetoric of Museums and Memorials*, ed. Greg Dickinson, Carole Blair, and Brian L. Ott (Tuscaloosa: University of
Alabama Press, 2010), 25. The authors render the relationship in a useful analogic
formula: "space : place :: time : memory."

4 Maurice Halbwachs, *On Collective Memory*, trans. Lewis A. Coser (Chicago,
IL: University of Chicago Press, 1992). See also Greg Dickinson, Carole Blair, and
Brian L. Ott, eds., *Places of Public Memory: The Rhetoric of Museums and Memorials* (Tuscaloosa: University of Alabama Press, 2010); Kristin Hass, *Carried to the
Wall: American Memory and the Vietnam Veterans Memorial* (Berkeley: University
of California Press, 1998); Marianne Hirsch, *Family Frames: Photography, Narrative, and Postmemory* (Cambridge: Harvard University Press, 1997); Alison Landsberg, *Prosthetic Memory: The Transformation of American Remembrance in the Age
of Mass Culture* (New York: Columbia University Press, 2004); Kendall R. Phillips,
ed., *Framing Public Memory* (Tuscaloosa: University of Alabama Press, 2004); and
Marita Sturken, *Tangled Memories: The Vietnam War, the AIDS Epidemic, and the
Politics of Remembering* (Berkeley: University of California Press, 1997).

5 Barbara Misztal, *Theories of Social Remembering* (Maidenhead: Open University Press, 2003), 6.

6 Susannah Radstone and Bill Schwarz, "Mapping Memory," in *Memory:
Histories, Theories, Debates*, ed. Susannah Radstone and Bill Schwarz (New York:
Fordham University Press, 2010), 2.

7 Kenneth Foote, *Shadowed Ground: America's Landscapes of Violence and
Tragedy*, 2nd ed. (Austin: University of Texas Press, 2003).

8 Achille Mbembe, "Necropolitics," trans. Libby Meintjes, *Public Culture* 15
(2003): 12, 11. For a more comprehensive analysis of the necropolitics of roadside
memorialization, see Robert M. Bednar, "Killing Memory: Roadside Memorials and
the Necropolitics of Affect," *Cultural Politics* 9, no. 3 (2013): 337–56. For an earlier
argument linking necropolitics to collective memory, see Gillian Rose, "Who Cares
for the Dead and How?: British Newspaper Reporting and the Bombings of London,
July 2005," *Geoforum* 40 (2009): 46–54.

9 Mbembe, "Necropolitics," 12.

10 For an accounting of the different state policies regarding roadside memorials,
see George E. Dickinson and Heath Hoffman, "Roadside Memorial Policies in the
United States," *Mortality* 15, no. 2 (2010): 154–67.

11 Mbembe, "Necropolitics," 27; emphasis in original.

12 Paul Williams, *Memorial Museums: The Global Rush to Commemorate
Atrocities* (Oxford: Berg, 2007), 48.

13 Erika Doss, *Memorial Mania: Public Feeling in America* (Chicago, IL: University of Chicago Press, 2010), 37. For a more recent work that places roadside
shrines in the context of similar memorial practices internationally, see Margaret
Holloway, Miraslava Hukelova, and Louis Bailey, *Remember Me: Memorialisation
in Contemporary Society* (Hull: University of Hull, 2018), accessed January 12, 2019,
https://remembermeproject.wordpress.com.

14 Doss, *Memorial Mania*, 2, 19. See also James Young, *The Texture of Memory:
Holocaust Memorials and Meaning* (New Haven, CT: Yale University Press, 1993).

15   See Karen Wells, "Melancholic Memorialisation: The Ethical Demands of Grievable Lives," in *Visuality/Materiality: Images, Objects and Practices*, ed. Gillian Rose and Divya P. Tolia-Kelly (Farnham: Ashgate, 2012), 153–69.

16   Pierre Nora, "Between Memory and History: *Les Lieux de Mémoire*," *Representations* 26 (1989): 7–24; Dickinson, Blair, and Ott, eds., *Places of Public Memory*.

17   Blair, Dickinson, and Ott, "Introduction," 29. For more on the politics of placed memories, see also Andreas Huyssen, *Present Pasts: Urban Palimpsests and the Politics of Memory* (Palo Alto, CA: Stanford University Press, 2003).

18   See Kirk Savage, *Monument Wars: Washington, D.C., the National Mall, and the Transformation of the Memorial Landscape* (Berkeley: University of California Press, 2009).

19   Blair, Dickinson, and Ott, "Introduction," 30; emphasis in original.

20   See Hass, *Carried to the Wall*. See also Karen A. Franck and Quentin Stevens, eds., *Loose Space: Possibility and Diversity in Urban Life* (London: Routledge, 2007); and Quentin Stevens and Karen A. Franck, *Memorials as Spaces of Engagement: Design, Uses and Meaning* (London: Routledge, 2016).

21   See Scott A. Sandage, "A Marble House Divided: The Lincoln Memorial, the Civil Rights Movement, and the Politics of Memory, 1939–1963," *Journal of American History* 80 (1993): 135–67. See also Owen J. Dwyer and Derek H. Alderman, eds., *Civil Rights Memorials and the Geography of Memory* (Chicago, IL: Center for American Places, 2008).

22   See Paliewicz, Nicholas S., "Bent but Not Broken: Remembering Vulnerability and Resiliency at the National September 11 Memorial Museum," *Southern Communication Journal* 82, no. 1 (2017): 1–14; Nicholas S. Paliewicz and Marouf Hasian, Jr., "Mourning Absences, Melancholic Commemoration, and the Contested Public Memories of the National September 11 Memorial and Museum," *Western Journal of Communication* 80, no. 2 (2016): 140–62; Nicholas S. Paliewicz and Marouf Hasian, Jr., "Popular Memory at Ground Zero: A Heterotopology of the National September 11 Memorial and Museum," *Popular Communication* 15, no. 1 (2017): 19–36; and Shari R. Veil, Timothy L. Sellnow, and Megan Heald, "Memorializing Crisis: The Oklahoma City National Memorial as Renewal Discourse," *Journal of Applied Communication Research* 39, no. 2 (2011): 164–83.

23   See Williams, *Memorial Museums*.

24   Doss, *Memorial Mania*, 94; and Williams, *Memorial Museums*. See also John Lennon and Malcolm Foley, *Dark Tourism: The Attraction of Death and Disaster* (Andover: Cengage, 2010); Richard Sharpley and Philip R. Stone, *The Darker Side of Travel: The Theory and Practice of Dark Tourism* (Bristol: Channel View Press, 2009); Craig Young and Duncan Light, "Interrogating Spaces of and for the Dead as 'Alternative Space': Cemeteries, Corpses and Sites of Dark Tourism," *International Review of Social Research* 6, no. 2 (2016): 61–72; and Phaedra C. Pezzullo, *Toxic Tourism: Rhetorics of Pollution, Travel, and Environmental Justice* (Tuscaloosa: University of Alabama Press, 2007).

25   Cf. Ekaterina Haskins and Michael Rancourt, "Accidental Tourists: Visiting Ephemeral War Memorials," *Memory Studies* (2016): 1–15.

26   Pierre Nora, "Between Memory and History: *Les Lieux de Mémoire*," *Representations* 26 (1989): 7.

27 Gregory Ulmer, "Traffic of the Spheres: Prototype for a Memorial," in *Car Crash Culture*, ed. Mikita Brottman (New York: Palgrave, 2001), 336.

28 Ibid.

29 Michel de Certeau, *The Practice of Everyday Life*, trans. Steven F. Rendall (Berkeley: University of California Press, 1984).

30 Ibid., xix.

31 Ibid.

32 Astrid Erll, *Memory in Culture*, trans. Sara B. Young (New York: Palgrave MacMillan, 2010), 9.

33 For a classic but still pertinent analysis of how the built environment of car culture addresses drivers, see Robert Venturi, Denise Scott Brown, and Seven Izenour, *Learning from Las Vegas: The Forgotten Symbolism of Architectural Form* (Cambridge: MIT Press, 1977). See also Lisa Mahar, *American Signs: Form and Meaning on Route 66* (New York: Monacelli Press, 2002).

34 I have also seen this orientation pattern in the United Kingdom and New Zealand, where the traffic flow pattern is the opposite of the traffic pattern in the United States. There, the orientation pattern is simply reversed to match the traffic flow, with road trauma shrines facing drivers driving on the left side of the road on a two-way road instead of on the right.

35 Jörg Beckmann, "Mobility and Safety," *Theory, Culture & Society* 21 (2004): 94.

36 Ibid., 94–95.

37 Mike Featherstone, "Automobilities: An Introduction," *Theory, Culture & Society* 21, no. 4–5 (2004): 3.

38 Beckmann, "Mobility and Safety," 97.

Chapter 4

Materializing Road Trauma

One of the defining characteristics of road trauma shrines is that they accumulate and magnetize stuff. After Daniel Miller, I use the term "stuff" purposefully here, even though it at first sounds informal or vague, as the word contains the distinction many material culture scholars argue over between "objects" and "things."[1] The stuff at shrines gets magnetized to other stuff in the shrine, pulling some stuff together and pushing other stuff away. And as shrines do this, they also work as magnetic forces in the roadside in general, where they attract attention to themselves, but also exist as polarizing figures in the culture.

Once a shrine has been established as a unique place, anything placed there becomes transformed through spatial and material practices that invest objects through transference and establish what Foucault calls "relations of proximity" among objects within the shrine and between the shrine and the surrounding landscape.[2] As we will see, even the most ordinary object is capable of being extraordinary, simultaneously containing traumatic affect while being a portal into a world. But at the same time, those other worlds are never as palpable as the objects that provide access to them, which complicates any analysis.

Kathleen Stewart argues, "Something huge and impersonal runs through things, but it's also mysteriously intimate and close at hand."[3] While they live their lives on the side of the road, trauma shrines generate what Stewart calls "the actual residue of people 'making something of things.'"[4] Every one of these things is entangled within a web made of both matter and imagination—of the real-and-imagined. As Ben Highmore puts it, "The sticky entanglements of substances and feelings, of matter and affect are central to our contact with the world."[5] Studying these entanglements can reveal "the way in which bodies, emotions, world trade and aesthetics, for instance,

89

interweave at the most everyday level."⁶ Thus, to analyze shrines as affective trauma places demands attuning simultaneously to the particular and the general at the same time, to see a shrine in its concrete form not only as a physical place but also as a location for collecting and magnetizing objects that traffic in the movement between the radically concrete and the radically intangible but somehow still intelligible and sensible.

As a specific type of trauma shrine, road trauma shrines are made of an array of material objects brought into relation with each other and with those who encounter them while being located in a unique space: at the site of a particular car crash, which is usually in the liminal space of the public right-of-way, a space of mobility used by many but "owned" by none. Road trauma shrines create place by collecting and magnetizing objects that come from "all over the map"—physically and culturally—that are brought to rest temporarily within a shrine. There, they do the work of giving an embodied and material form to the identity of the trauma victims—both the victims of the bodily trauma who died at the site and the "survivors" of the crash victim who build and engage with the shrines—so that the sites can do the work of managing trauma and grief in place, over time, for intimates while also addressing a much wider public.

The evolving life of a road trauma shrine is predicated on a dynamic that transforms the shrine into a prosthetic or proxy for the lost body it represents. Such a transformation depends of some force or relation of forces capable of making ordinary objects and spaces extraordinary, and some force or relation of forces capable of locating the shrine in a particular place where it is always in place and out of place at the same time. As we will see, there are two main dynamics at work here. At the scale of individual objects, the force is *transference*: the investment of material objects with immaterial significance through cultural practices. At the scale of relations among objects in the shrine and between the shrine and its immediate surroundings, the force is more a *relation of forces*: the relations that make road trauma shrines function as what Foucault calls a *heterotopia*. Let's look at each of these forces in turn before we look at how they work in the field.

TRANSFERRED LIVES

Without some dynamic process of investing material objects with a connection to the lost person, a road trauma shrine simply would not work. That dynamic process is transference. Transference is a key concept in psychoanalytic theory, first in Freud but also in more contemporary psychoanalytic approaches, particularly Object Relations theory. The concept is used to analyze how people project their internal conceptions of the world onto particular external objects—people as well as things—within intersubjective encounters. Transference is especially complicated to trace in a therapeutic relationship, where there is not only transference from the client to the therapist but also counter-transference from the therapist to the client. In more traditional psychoanalytic approaches, transference was seen either as a fantasy projection that often prevented patients from seeing things as they were or as a therapeutically useful form of fantasy that could be deployed to help people work through neuroses. Contemporary approaches aligned with poststructural theory treat transference more as an intrinsic and inescapable part of *any* relationship, not only therapeutic relationships, because all identity is performative and each person in a relationship mutually constitutes each other—making the other an object of their projected desires and fears, and vice versa—instead of seeing transference as a pathological projection of a false conception that must be corrected.

Also in recent years, scholars in the humanities and social sciences have been applying this psychoanalytic concept to a more literal kind of "object relations" between people and physical objects within material culture. With transference, the subject treats the object as if the object is equivalent to the subject's *idea of* the object, erasing difference through "projective identification." Projective identification is, as Peter Redman argues, "a process of unconscious communication firmly located in the present and within a relational field," where transference dynamics "are simultaneously internal and shared, felt as belonging inside a particular individual while having no clear home in any single person."[7] The dynamic of transference is radically uncontained—"a flow rather than a location" where "the 'inner' always has the 'outer' present within it (and vice versa) such that the boundaries between inside and outside are fundamentally blurred and unstable."[8]

This aligns the concept squarely with the model of individual and collective trauma and memory I am outlining here, where both memory and trauma are performative, and thus always experienced in the present, even when they are "about" the past, and even when the content of a traumatic memory cannot be directly shared through a representational economy. Crash shrines materially assert the presence of lost objects, but do so not by literally dwelling *in* the past—which would entail somehow living in the past—but by dwelling *on* the past by performing memory in the present. As in the therapeutic treatment of trauma, which might seek to make repressed or dissociated memories present so as to work through them in the space of therapy, the challenge for a shrine is to bring traumatic loss into the light of day, where it can be worked on. Therefore, each crash shrine thus materializes not only the trauma of crash victims but also its own history as a site of working-through. In some shrines, the transference is even more literal, where a plant is transplanted at the site—where, if cared for long enough to be established, the growing plant will take the place of the person whose life is now separated from time (see figure 4.1). This indicates that the work of transference at shrines is ongoing, and compensatory—happening as long as the site is actively used.

Transference is not a matter of simply *placing* or *investing* meaning in particular objects in a particular physical location, but an ongoing *practice* of performative engagement with the visual, the material, and the spatial, and a way of being-in-the-world that connects the material to the intangible and vice versa. As Gillian Rose and Divya P. Tolia-Kelly observe, "Practice is what humans do with things."[9] Transference is a *doing*, a *practice*. Road trauma shrines are things and collections of things, but they are also a location of ongoing practices of transference, where road trauma shrines transfer the lives lost in car crashes to the lives lived by shrines as assemblages as well as the memory objects and spaces contained within shrines.

In short, road trauma shrines produce affect through transference to keep the memory of absent people present in the everyday landscape. Placing objects in shrines makes everything at the shrine site a transference object so that the shrine comes to embody the person or person it commemorates in a material as well as metaphorical way. There on the roadside, a shrine keeps absent people present by materializing objects that produce evidence not strictly *of a loss* but *what was lost*: a life interrupted—a life being lived, with a taken-for-granted future and a unique identity and history, which was violently interrupted by a car crash—but also a *relationship* interrupted, or more precisely, transformed. The person remembered is gone, but proxy materializations of them abound in the shrine, and these objects become the location for the performance of continuing bonds.

If shrines transfer the life of the victim to the shrine, ensuring that the shrine will live past the victim, shrines not only literally *re-place* the victim in social space but also *compensate* for the victim's lost future as a social entity. While the victim will no longer celebrate new birthdays, anniversaries, and holidays, *the shrine does*. The shrine as living memory/space is a means of replacing the interrupted life of the victim with a new life that is allowed to take its course as it was expected to before it was interrupted by the crash. This applies not only to the shrine's life but also to its death. While the victim's life was cut short, without the possibility to live until the "natural" processes of bodily decay prevail, the shrine is allowed that privilege, too: the privilege of spending its everyday life standing there in the wind and the sun and the rain, doing its job, like a person, living its life instead of having its life cut short by a tragic and untimely death.

Thus, one of the things shrines stage is a recreation of the lost body's ability to live long enough to die of natural causes. Sometimes this fact is so evident that it is uncanny, as it is in a shrine north of Albuquerque, New Mexico (see figure 4.1). The site has featured a stuffed cartoon Tasmanian

Figure 4.1. Top: *Road Trauma Shrine with Live Fir Tree*, US-285-South, Pojoaque, NM, USA, February 2010 and July 2019. Bottom: *Aging Tasmanian Devil*, Interstate Highway 25-North, North of Albuquerque, NM, USA, March 2006 and February 2012.

devil at least since I took the first picture in 2006, where he appears new even as the photo in the frame above "TAZ" is faded beyond recognition, indicating a site history even longer than I picture here. In 2012, I photographed the site again and "TAZ" has aged considerably, including graying eyebrows. Clearly, the iconic and anthropomorphic aspects of these objects contribute to the uncanny feeling that you are witnessing a transference. We will explore the performative dimensions of this process of prosthetic "aging" more fully in chapters 5 and 7.

Once that first transference of absent body to present object is materialized in a shrine, these places then act as a proxy for the absent victim as the shrine takes on a life of its own in the public right-of-way, where each shrine sets up a space of witnessing and contributes to a collective remembrance of the everyday traumas enmeshed in automobility. Clearly, there is a spiritual and theological dimension to this, and different shrines perform different conceptions of the afterlife. However, all seem to materialize a belief that a shrine is a portal through which the living can communicate with the dead as if they are still present. And, as we will see, that process of *making-present* and then *keeping-present* is the key to how they work at both the micro and macroscales at once.

HETEROCHRONIC HETEROTOPIAS

Transference works to invest the objects in shrines with the presence of absent bodies. However, if there were not some other, larger-scale relation of forces, transference would have only minor effects. That larger-scale relation of forces is what I would like to address now. Shrines materialize and activate traumatic rememories that might otherwise be contained only in the bodies of those who knew the victims but never find a material or spatial form, much less a social presence. And they do so in a physical and discursive space—the roadside—where they create the potential for a public memory where there otherwise would not be one, materializing rememory not just for those who know the people being memorialized but also for strangers as well. Crash shrines commemorate the loss of individual lives, but because they do so within a location discursively produced as individually shared collective space, they also perform a loss to the collective.

But while they perform memory of loss, they also assert its opposite: continued presence. A shrine is not only a living memory of someone lost—an active attempt to keep the memory present and alive in the public sphere—but also a talking-back to that death itself through material means, an assertion not only of the memory of absence but also of presence: *they are still here, included socially in the motoring public driving by them every day*. In short, road trauma shrines create places for remembering individual road trauma that assert an ongoing social presence of lost drivers and passengers.

As such, road trauma shrines are places that show and perform trauma for both inside and outside witnesses through experiential mirroring. They do so because they are structured as what Foucault calls *heterotopias*. Foucault introduced the term in a 1967 lecture first translated into English in 1986 as "Of Other Spaces," and in 1988 as "Different Spaces."[10] Foucault defines a heterotopia as a "counter-site" that reflects, refracts, and inflects surrounding social spaces externally while internally creating "relations of proximity" between spatial practices and objects otherwise marked and policed as separate: public/private, sacred/profane, global/local, material/immaterial, mediated/personal. While heterotopias create *spatial* relations of proximity among objects within a heterotopia and between a heterotopia and its surroundings, heterotopias also create *temporal* relations of proximity between and among times: now/then/and then, not only here/there. Therefore, a heterotopia brings together not only multiple *spatialities* but also multiple *temporalities* that jar against one another as well, creating a space of material and cultural simultaneity and juxtaposition. That's why it would be best to call these forms *heterochronic heterotopias*: they are *other spaces* that function as heterotopias exactly because they also contain and reflect/refract *other times*.

Foucault's concept of the heterotopia has been engaged within diverse analytical projects in British and American cultural studies as well as in fields as diverse as literary studies, sociology, and geography.[11] Here I want to build on that work by showing that Foucault's concept of the heterotopia is particularly useful for theorizing phenomena within automobility, where spaces, bodies, objects, texts, and discourses converge and collide through continually shifting spatial, temporal, and cultural relations. It's not a concept that theorizes the relations within a larger cultural apparatus such as automobility, however. Indeed, although much of Foucault's work is about large-scale regimes and discourses, his work on heterotopias makes no mention of a system or network of relations *between or among* distinct heterotopic spaces, only *within* heterotopic spaces, or between heterotopic spaces and the spaces surrounding them.

To me, the concreteness and rootedness of a heterotopia is the main appeal of the concept as I have sought to understand how road trauma shrines work in particular locations. As a point of comparison, a term that also captures the sense of immanence and becoming as well as the contingent relationality of the heterochronic heterotopia is Deleuze's concept of "assemblage."[12] For Deleuze, assemblage is both a verb and a noun, meaning that the process of *assembling* heterogeneous elements together creates at least a temporary *assemblage* of those elements. As such, assemblage as both process and object happens and exists at all scales, from "inside" bodies to between and among bodies. That makes it a useful concept to explain all sorts of momentary, rhizomatic, and nonteleological, and nonessential real-and-imagined relations of proximity, but the same is true of the heterotopia while the heterotopia concept also helps us understand how such shifting relations and the things they collect and magnetize are also sometimes uniquely placed in space and time at radically unique places.

However, as we try to theorize the *interrelationships among* different shrines at different locations, which somehow stay separate as they cohere together to evidence a collective, cultural trauma, the assemblage concept is more useful than heterotopia. A single road trauma shrine is made up of what Jane Bennett calls "vibrant matter" and is always more than the sum of its parts. But the same is true of roadside shrines taken as a dispersed but related *assemblage,* just at a different scale. As Jane Bennett argues, "Each member and proto-member of the assemblage has a certain vital force, but there is also an affectivity proper to the grouping as such: an agency *of* the assemblage."[13] This is something also worked out by Manuel DeLanda, who especially emphasizes that the relations within an assemblage are always temporary "relations of exteriority" instead of essential or transcendent "relations of interiority"; that is, in assembling things in an assemblage, the relations are built from "capacities" and "emergences," not static or essential functions and meanings.[14] As I argue throughout the book, each road trauma shrine has a radically unique agency in the cultural landscape derived from how it works as a heterotopia comprised of transference objects magnetized together within certain sites. Taken together as a contingent *assemblage* of heterotopias instead of an always-already coherent and bounded *whole,* shrines form a loose and nervous collective that has an agency all its own.

In keeping with the localized scale of the heterotopia concept, and in keeping with the aims of the book as a whole, however, I want to show how the concept of the heterochronic heterotopia helps us analyze road trauma shrines as radically unique, concrete places where intangible things are also made present. Roadside shrines to car accident victims create real-and-imagined in-between "other spaces" that produce temporary relations of simultaneous proximity and exceed their boundaries—neither staying in one place nor representing one time, continually absorbing new objects and practices and dispersing others—all on the equally heterotopic space of the public right-of-way, where bodies, objects, machines, and institutions temporarily inhabit the same roadspace at the same time using a common cultural resource for divergent ends. In short, in producing different spaces through the composition of different elements, they create an *other space* that adds a different dimension to spaces of automobility.

But if, as a heterotopia, a road trauma shrine is an "Other Space" or "Different Space," the question is: Other than what? Different from what? In an article on crash shrines that also refers to Foucault's heterotopia, Catherine Collins and Alexandra Opie argue that roadside shrines are "Other Spaces" in the sense that they provide a space for working-through the violent deaths that occur there by creating an orderly inversion of "the chaos of traumatic memory and grief."[15] I see this occurring as well, but I want to emphasize that shrines also function as "other" and "different" in more spatial ways as well, which range from having the effect of mirroring as well as inverting. Whatever shrines do for the people who build them, they also reflect, refract, and invert the physical and cultural spaces they inhabit as well.

Just as shrines produce space, they also produce time. And like other heterotopias, they have a complicated relationship to time. In "Other Spaces,"

Foucault makes a distinction between "heterotopias of indefinitely accumulating time," such as museums and libraries, and heterotopias that mark a sharp break in the flow of time, such as cemeteries. Looking at roadside shrines, you can see elements of both relationships to time: a shrine not only marks the decisive break between life and death but also insists that the break be performed in the present and in public, where it must be encountered regardless of a witness's desire, in a different kind of indefinitely accumulating time. Also, unlike the cemetery, museum, and the library, which are already institutionally produced as a structured space of storage and memory, where bodies or objects are institutionally classified into predetermined and relational plots, wings, or locations, a roadside shrine is always a one-off, singular production that has to establish its own rules of engagement as it produces place. There are of course implicit vernacular cultural rules for how shrines should be built and where, but that is a much different type of organization of forces than corporate or government policy.

Which brings me to my final point about the way road trauma shrines work as heterochronic heterotopias. In his critique of the legacy of Foucault's heterotopia concept, sociologist Peter Johnson notes that the word heterotopia "is originally a medical term referring to a particular tissue that develops at a place other than is usual. The tissue is not diseased or particularly dangerous but merely placed elsewhere, a dislocation."[16] The point Johnson emphasizes is also the one I emphasize here: that the key fact about heterotopias is their material and represented relations of difference both placed and displaced within the sites and between the sites and their surroundings—not the *value* of that difference, particularly the cultural politics of such a site. Johnson argues that although scholarly uses of the heterotopia concept—especially within cultural studies—often have had a "persistent association with spaces of resistance and transgression," it would be wrong here to say that a heterotopia necessarily is a utopic space—a perfect or perfected space or even a necessarily progressive or resistant space.[17]

As we saw in chapter 3, each road trauma shrine defines itself as a place tactically by marking its difference from the surrounding space while also bringing together different spaces and times within it, but it does so in a cultural situation where it cannot define all of its terms and where it does not control the overall space within which it is performed. In this, it seems to me to be the perfect heterotopic mirror to reveal the structure of automobility in the contemporary United States, where each individual is simultaneously subject to vast and deep organizational systems that produce drivers who share the road with other drivers sharing the road, where all of the drivers drive their own cars to their own destinations, for different purposes and with different motives and to different effects—where they drive alongside each other literally and figuratively, together, if only for a moment—and where they also sometimes drive *into* each other. Each road trauma shrine simultaneously is this structure and shows this structure at the same time.

HOW ROADSIDE SHRINES MATERIALIZE
ROAD TRAUMA

Every road trauma shrine is a tactical heterotopia built from and working through transference objects tactically placed at a unique location within the roadscape. While recognizing the limits of linguistic metaphors when speaking of material culture, it is worth saying at the outset that road trauma shrines have a clear paradigmatic vocabulary and syntactical grammar of dominant objects and object relations. That is, there are certain objects that are prevalent at roadside shrines, and those certain objects are often grouped together in patterned ways as well. Placing these objects at the site is itself a performative act, and once the objects are placed at the site, they become the medium through which other performative acts are materialized. We will touch on this temporal process just a little here, but consider it much more fully in chapter 5, which is less about the *things* placed at shrines as objects than *what happens to and through them* once they are placed there.

Material transference objects come in three main forms: *Offerings*, *Relics*, and *Portraits*. *Offerings* are objects brought to the site by those who build and contribute to them. These are things like crosses, flowers, teddy bears, and other mass-produced objects that circulate elsewhere in the culture in multiple forms and potentially could mean and do a number of things in other settings but are put to work in very specific ways in shrines in general and in even more specific ways alongside other transference objects at particular shrine sites. Some sites also include a number of *Relics*: much rarer, existentially unique things that have particular material, indexical, and/or embodied relations to the people memorialized in the crash or the people doing the memorialization, such as vehicle debris gathered at the crash site, articles of clothing work by the victim or by grieving subjects, and handmade objects. Photographic *Portraits* of accident victims are a kind of relic, in that they have indexical and iconic relations to the bodies they represent, but because they function in even more complex ways at these sites, I will consider them separately from other relics. To show how the placement of objects works at road trauma shrines, we will look closely first at examples of the three broad types of objects—Offerings, Relics, and Portraits—before exploring how objects are placed *together* into heterotopic and heterochronic *Relations of Proximity* with other objects to form distinctive compositions and relations among shrine objects.

OFFERINGS

In this section, we will look at the primary building blocks of road trauma shrines: the material objects mourners place there either when the site is built the first time or revised as a whole site, or when a mourner makes a pilgrimage to the site. When these objects are placed in relation to other material objects, especially the dominant central vertical element—usually a cross, but also often some kind of platform such as a grotto, sign, memorial stone—they make up the main materialization of trauma and grief of any road trauma shrine.

One of the most obvious offerings at road trauma shrines in the United States is some version of a Christian cross—more technically called a Roman or Latin cross, a cross that is taller than it is wide. Such a cross (or group of symmetrically sized crosses, if there are multiple victims, where each cross serves as a proxy that "stands for" a victim) is usually placed in the center of the shrine and thus serves to physically as well as discursively anchor the site: it not only gives structure to the site but also serves as a physical and metaphorical *platform* for the whole shrine. More important, over time, a cross serving as a platform for other Offerings on the roadside has become the single most important material marker that communicates the identity of the cultural form at speed and distance: *This is a roadside car crash shrine.*

Of course, not all shrines have a cross as their central vertical element, but ones that don't usually do not have any alternate religious symbol instead. Indeed, in my years of fieldwork in the American Southwest and even in my more incidental travels throughout North America and Central America, I have only witnessed two roadside shrines that had a Star of David instead of a cross as the central, recognizably religious element of their design: one near Taos, New Mexico, and one on the western shore of the Salton Sea in California (see figure 4.2). Other common platforms for shrines are grottoes (derived from other roadside *capillas* and yard devotional shrines in Roman Catholicism, and more prevalent in the Borderlands), chairs, and carved or fabricated memorial stones like the ones you might otherwise see in a cemetery (see figures 4.3–4.4).[18] As we saw in the previous chapter, many shrines of the Roadside Attachment variety actually use a piece of the roadside infrastructure as their central vertical element. In general, only Roadside Island shrines actually *need* to have a central vertical element to hold them together and make them recognizable as roadside shrines.

The overwhelming predominance of crosses at road trauma shrines is striking, but it would be wrong to conclude that a cross's presence necessarily signals that victims or their families are Christians and that they have certain specific sectarian beliefs. It is even difficult to prove based on the material

Figure 4.2. Left: U.S. Highway 64-East, West of Taos, NM, USA, August 2003. Right: California State Highway 86-South, Salton City, CA, USA, July 2006.

Figure 4.3. Left: Arizona State Highway 85-North, Ajo, AZ, USA, August 2006. Right: U.S. Highway 89-North, North of Flagstaff, AZ, USA, August 2006.

Figure 4.4. Left: Arizona State Highway 85 @ Arizona State Highway 86, Why, AZ, USA, August 2006. Right: Parmer Lane @ Texas State Highway 130, Austin, TX, USA, February 2008.

evidence of a shrine that anyone associated with a shrine is affiliated with a particular faith tradition simply by looking at the shrine. We can make such an assumption in more closed social environments like Northern New Mexico and South Texas, where roadside shrines and cemetery gravesites also are decorated almost identically, and it is safe to assume that Latinx individuals memorialized with the iconography of Latin Catholic folk traditions are probably Roman Catholic. Certainly, there also is plenty of evidence that mourners articulate their grief in keeping with a Christian cosmology, where victims are said to "go straight to heaven" when they die in crashes. However, I think it is worth noticing that these expressions of faith in an afterlife are usually so generic as to be thought of as expressions of the kind of popular cultural Christianity that circulates in American dominant culture more than demonstrations of religious affiliation. That is, in a predominantly Christian nation, where Christianity is the privileged, naturalized, and presumed norm, symbols like the cross and angels are normative visual resources used to mediate grief and trauma because they reflect dominant cultural beliefs and practices about death and mourning, not only the elaborate theological beliefs forming and informing them.[19]

Figure 4.5. New Mexico State Highway 518-North, South of Taos, NM, USA, December 2010.

The complexity of this question is reflected in the mixed record of court cases where people and organizations have sued to have roadside shrines and other memorials that use religious iconography removed from the public right-of-way because they appear to promote religion in a nation founded on the belief in freedom of religion and separation of church and state. Critics of roadside crosses are right to argue that the cross will always represent Christianity even when it is being used more generically to represent "death." Just as the popular phrase "Rest in Peace" or RIP has its roots in Christianity but is cited by non-Christians, every citation of RIP or the cross carries with it that original association with Christianity. Either way, what is clear is that the prevalence of crosses indicates that crosses are an implicitly naturalized and normative element of the visual and material vocabulary of grief and mourning in contemporary American popular visual and material culture.

Regardless of their intended or functional association with Christianity, the other important thing about crosses incorporated into roadside shrines is that they are perfectly shaped to serve as transference prosthetics for the people they are used to memorialize. The cross maps almost exactly to the human body: the top vertical section of the cross figures the head, the two horizontal sections figure the arms, and the bottom vertical section figures the torso and legs. The cross is thus usually the most direct way that the victim's interrupted life is both materialized and figured as a proxy in the life of the shrine.

This is most obvious when the cross's "head" features a close-up photograph of a victim, but also when the "neck" or "arms" of the cross support jewelry as if it is not simply hanging on a pole but being *worn* by the cross.

Especially in the early days and weeks after a crash, mourners make what amounts to a pilgrimage to the roadside shrines, either individually or in groups. Some people will prepare for the visit in advance, either making or buying things and bringing them with them to place at the shrine. These objects are what I call "premeditated offerings." The stuff of premeditated offerings is the dominant material culture of bereavement in the United States today: things like flowers, stuffed animals, balloons, angels, candles, and popular religious and folk or commercial bereavement art. These objects reflect a deeply cultural agreement about what is appropriate to bring to any bereavement site. You will see them all well represented in the photographs of shrines throughout the book, so I will wait to analyze them until we discuss relations of proximity later in the chapter.

However, this is not the only kind of offering you see at road trauma shrines. Indeed, while many shrines feature premeditated offerings, and many people go to shrines outfitted with offerings to place there, it is clear that other people must decide to leave things there after they get there, because there are also all kinds of objects at roadside shrines that are not necessarily associated with bereavement or (presumably) the identity of the bereaved. These are what I call "participatory offerings" instead of premeditated offerings; they are the kinds of things people probably leave as offerings because they want to participate in the practice of placing an offering but didn't bring any bereavement objects with them. That means that they are usually objects that the pilgrims had "on their person" or, more precisely and significantly, *in their car*, which get transformed into an offering through a recontextualization of the object as a bereavement gift: wristbands, lanyards, jewelry, pens, buttons, key chains, coins, cigarettes and cigarette lighters, etc.—even a toothbrush and a calculator in one case.

Participatory offerings provide us with direct evidence of two phenomena central to material culture studies: singularization and recontextualization. Both of these concepts draw attention to what Arjun Appadurai calls the "social life of things" and Igor Kopytoff calls the "cultural biography of things": the way objects change meaning and both gather and disperse affective performative power as they are moved from one physical, cultural, and/or discursive context to another.[20] All objects can be recontextualized as they move in the world, from the most unique handmade and singular objects to the most common mass-produced objects. As they move and are moved from context to context, they become discursively produced in different ways. As a specific form of recontextualization, singularization refers to the process experienced particularly by mass-produced objects, where one particular commodity out of many is singularized by being chosen out of a horde of identical objects on the shelf and given unique meaning. To singularize a mass-produced commodity, it first must be decommodified (purchased and given a singular purpose) and then it must be placed somewhere where it is associated with a singular person or group, where the object gets tied up with their identity.

Both singularization and recontextualization are apparent in the case of participatory offerings, where objects both intended for and often associated with particular uses are placed in a road trauma shrine, and where they are made to serve much different purposes. That is the source of their poignance in most cases: where a toothbrush in a shrine resonates in a totally different way than it does on a bathroom counter, where a rearview mirror at the base of a shrine feels very different than one on the door of a car, and where a tree in a shrine feels different than the ones just a few feet away in the forest. Of course, we should say that the same thing happens to more standard bereavement objects within a smaller range of recontextualizations. There are much fewer options for how a silk funereal "Mom" flower arrangement can be used, but teddy bears can be toys, cure expressions of love, or comfort objects for grieving, just as fresh-cut flowers are given to express both affection and grief.

Figure 4.6. Left: Arizona State Highway 86-East, Three Points, AZ, USA, August 2006. Top Right: Arizona State Highway 86-East, Three Points, AZ, USA, August 2006. Bottom Right: New Mexico State Highway 68-West @ Rio Grande Gorge Bridge, West of Taos, NM, USA, February 2012.

Here we should also note that there are specific ritual practices associated with roadside crash shrine pilgrimages and participatory offerings. The most intriguing of these is the practice of bringing food and drink to the site to share with the victim. The most prevalent version of this is the practice of raising a toast to the victim, either by gathering to drink the bereaved person's favorite alcoholic drink in honor of the victim, who can no longer enjoy such a thing, or by more literally sharing alcohol with the victim, either by leaving unopened beer or liquor bottles and cans at the shrine, or by pouring them on the ground at the site.

Because these offerings were carried "on the person" of the mourner when they decided to offer them to the site, they could also be seen as relics, but we will focus on them in this section because they function mainly as offerings, as gifts offered as a symbolic "tribute" to the dead. We will analyze a final type of relic offering, handwritten messages addressed to the victim inscribed on other Offerings, Relics, and Portraits, in chapter 5. Finally, an exception to my distinction between premeditated offerings and participatory offerings is present at shrines for military personnel, members of sports teams, or strong subcultures such as Harley-riding bikers, where pilgrims bring pieces of their shared gear, or the material culture associated with the subculture, which expands the range of dominant premeditated offerings at these sites.

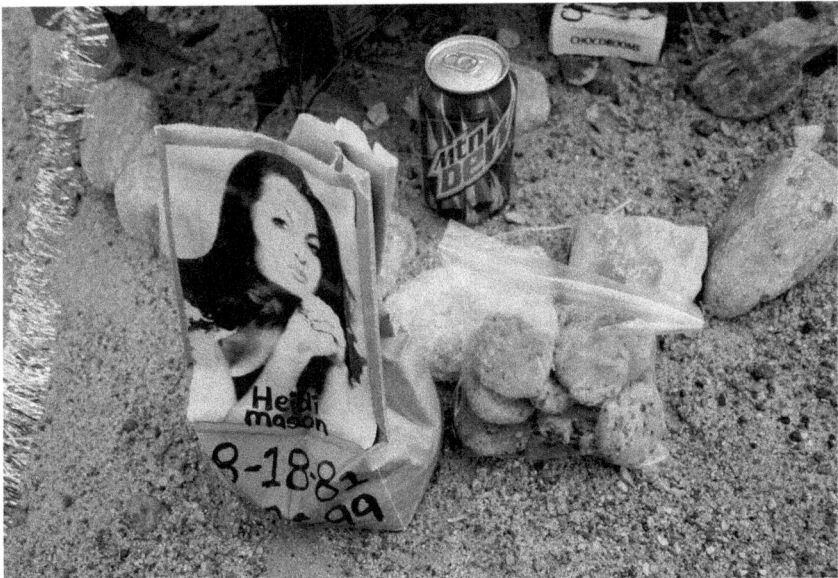

Figure 4.7. University Boulevard @ Rio Bravo Boulevard, Albuquerque, NM, USA, February 2012.

RELICS

A relic is something that has an embodied, indexical connection to some past life, an object that was physically associated with someone or something from the past, but that exists now in the present. A relic often not only is "from the past" but also makes that anachronistic status materially self-evident by being obviously not only *out of place* but also *out of time* in its current context. In the case of road trauma shrines, a relic is any material object that was touched by the bodies of the people associated with the shrine. By that definition, we might include everything at a shrine, because everything there was placed there by some *body*. However, what I mean by relic here is any material object that is associated with the intimate, everyday world of the crash victims and the people who knew and loved them before they perished in the crash. Strangers sometimes place participatory offerings at roadside shrines as well, but this is much rarer than intimates leaving offerings. All of these objects function as what Margaret Gibson calls "objects of the dead," but do so out on the roadside.[21] There, unlike other physical objects that function as relics and thus might be cherished as valued keepsakes, these objects are outdoors, in the public right-of way, where they are subject to disappearance, theft, and decay.

Within this definition of relics, there are basically three different types of relics. First, there are objects associated with the life of the victim before the crash, which belong to the world before the trauma the shrines mediate, such as articles of clothing and photographs of the victims. The second and to me most resonant type of relics located at road trauma shrines is an object that evidences the crash—pieces of crashed vehicles found on the site and incorporated into the shrine. As we saw in chapter 3, sometimes crash debris is simply "left over" from the crash and scattered about the site, but other times, the debris is gathered into the shrine, where the shrine places and "cites" the material from original crash to make it materially self-evident that a road trauma shrine mediates a road death, and to materially recontextualize it through the ongoing grief work happening at the site. Both of these types of shrine relics come into the present from a different time and place: from either before or during the crash. The third type of relic is produced entirely after the crash, when mourners leave objects tied to their own bodies as offerings to the victim. These relics are much more directly about performing a continuing bond between grieving subjects, the lost victim, and the other materializations at the site.

To show how this works, I would like to look in detail first at a particular shrine that contains all three forms of relics in one site. The shrine is for two members of the Texas A&M University Corps of Cadets who were killed in 2009 in a fiery crash north of Bastrop, Texas. The shrine itself is a Roadside

Island anchored by a mounted piece of plywood. Two small Texas A&M flags fly from the top of the shrine, making their university identity and affiliation clear even to strangers driving by.[22] On the plywood are messages written in Sharpie along with a number of relics. The writing includes two Bible verse numbers and identifies the victims as Tanner Ferris and Shawn Shearod of the Corps Company B "Street Fighters," graduating class of 2013, meaning that the men had just joined the Corps their freshman year the previous fall. The relics are from either the crash victims or the other Corps members who built the site: golden Corps pins, Corps uniform hats, Company B stickers, sheet music for the Corps' signature marching song attached to a horn clip. In this case, the person who built the site, another cadet named Will Bird, has identified himself, which makes the handwritten note from Bird itself a kind of relic as well.

At the base of the shrine and scattered all around it are different relics: the debris left over from the crash itself and the significant emergency effort required to keep the fire caused by the crash from spreading to the nearby pine forest in an area beset by massive forest fires in recent years. Taken together, these different relics provide extensive material evidence of a number of different traumas—the crash, the rescue effort, and the subsequent trauma of the people left behind to deal with both—all of which make the site heavy with affect.

Figure 4.8. Texas State Highway 21-South at Travis Road, North of Bastrop, TX, USA, July 2010.

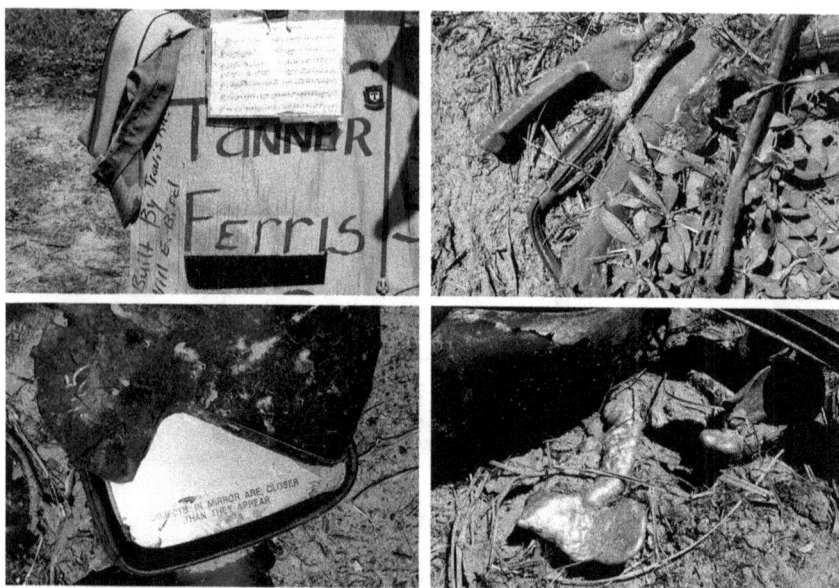

Figure 4.9. Texas State Highway 21-South at Travis Road, North of Bastrop, TX, USA, July 2010.

More often, shrines include relics only in the shrine itself. When indexical objects like hats and other pieces of clothing are combined with other relic offerings that are obviously connected in some existential way to either the victim or intimate mourners of the victim, the site resonates with the materialization of many affective performative acts. Further, when those relics themselves stay in place at the shrine long enough to age on the roadside, the transferential connection of body to relic is performed over time, simultaneously making the person present in a palpable way, but also absent. This is most visible for me at shrines containing clothing worn by the victim, but also where there is any kind of clothing or jewelry or "gear" present. These objects that had once been part of someone's everyday embodiment are no longer functioning as they once were before the crash but as a relic offering aging on the roadside without a body to hold them up. You may recall that this was the case also at the site where we opened the book, where the hockey team portrait is placed next to the hockey stick he is holding in the portrait itself, collapsing the different places and times of the boy's life and death into a kind of heterotopic and heterochronic vortex of materialized affect.

When relics from the crash itself are incorporated into the shrine, it not only intensifies the material communication of the trauma at the site but also gives it a certain inflection. I see these instances as a kind of citationality,

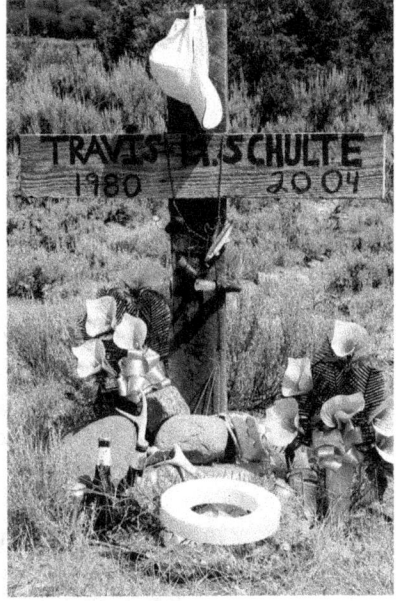

Figure 4.10. Left: U.S. Highway 290-East, Elgin, TX, USA, March 2015. Right: Colorado State Highway 149-South, North of Lake City, CO, USA, July 2006.

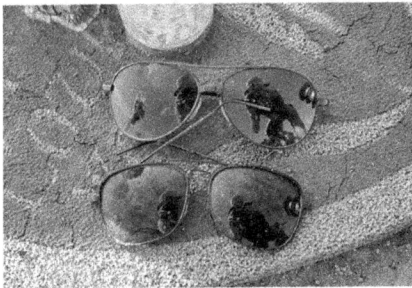

Figure 4.11. Top Left: Old Santa Fe Trail, Santa Fe, NM, USA, July 2006. Top Right: Kearny Villa Road @ Gun Club Road (Edge of Miramar Marine Corps Air Station), Mira Mesa, CA, USA, July 2006. Bottom Left: Arizona State Highway 86-East, Three Points, AZ, USA, August 2006. Bottom Right: E. Palmdale Boulevard @ 150th Street-East, Palmdale, CA, USA, July 2006.

where the shrine "cites" the crash—bringing it from one physical and discursive reality to another, where it can be worked on—the way that citations do in writing and in other cultural productions today, where citationality is the norm.[23]

The citation of crash debris makes these shrines qualitatively different from most road trauma shrines, especially the ones with photographic portrait relics in them. Roadside shrines that do not include crash debris are more clearly focused on implicitly talking back to the crash by showing the lost victim as they were *before* the crash instead of focusing on the crash itself, which serves to stage the ongoing presence of the person lost, *as they were*. But here, there is a materialized tension between that desire to celebrate a life lost and the actual evidence of the death in the shrine—where the crash itself is palpably present even as the rest of the shrine focuses on the person lost in the crash and the grief of mourners who have taken on the work of keeping their memory alive through continuing bonds. This makes these sites much more complex in their affect and effects, and when a relic from the crash is made into the main anchoring object of the shrine, it provides an even more powerful material performance of this tension.

There is another intriguing pattern about this practice of citing the crash debris: it is more common in shrines devoted to victims of motorcycle accidents than any other kind of crash, especially those that display an affiliation with Harley-Davidson subculture. The practice reinforces the emphasis within the subculture on the rebellious, devil-may-care image: a kind of materialization of the "Ride Free or Die" or "Live by the bike, Die by the bike" attitude of many bikers. Riding a motorcycle also involves a more intimate hybrid of rider/vehicle than a car ever does so that the bike feels more like part of the rider and vice versa. That is, unlike other car crash victims, whose celebrated "true" lives may be located far from the road where they were killed, bikers are people whose lives are always entangled with the road and the bikes they use to ride it, so citing the crash debris in a shrine about a biker's life before the crash is one final performance of that affiliation.

PORTRAITS

Now that we have established how shrines make place and materialize objects, I want to turn to a special kind of relic: the photographic portraits that people use within roadside shrines. Because they are photographs, it might be easy to see them only as representations of the dead, but analyzing them here in the context of other materializations of trauma, memory, and grief will make it clear that shrine portraits function as visually and tactilely signifying *objects* that are more than visual, more than material, and thus

Figure 4.12. Arizona State Highway 86-East, West of Tucson, AZ, USA, August 2006.

visual/material. In the next chapter, we will explore how these visual/material objects have lives separate from but connected to the things they represent, and thus a particular form of performative agency, but the focus here will be on them as objects that are a primary force for anchoring and materializing trauma at roadside shrines.

What does a photograph do for a road trauma shrine, and what does a shrine do for a photograph? More specifically: What does displaying a photograph in a roadside car crash shrine do to the photograph's relationship to time and space, to life and death, and to the viewer, the producer, and the subject of the photograph? Photographs are powerful as representational objects because they are simultaneously iconic and indexical. That is, using the terms developed by Charles Sanders Peirce, photographs are both icons and indexes. They not only *resemble* the objects they picture (icon) but also *have some material connection to* them (index). Photographs look like the objects, places, and people they represent because those things were materially present in front of the camera at the time the photograph was made. We know this, and build it into the way we interact with photographs—when we pose for them, when we take them, when we display them, when we look at them, and when we talk about them. When photographs from the past find us in the present, they do so therefore not only as *depictions* of the past but also as a kind of *relic* from the past that is now present. As Kathleen Stewart writes in *Ordinary Affects* (2007), when we feel the power of photographs moving or touching us, the most profound affect we are witnessing is "the profound experience of watching images touch matter" and bringing it into the present.[24]

In *Camera Lucida* (1981), Roland Barthes argues that every act of viewing a photograph is a potential encounter with its "intractable" relationship to some material reality that has or has had an existence independent of the photograph: "Photography never lies: or rather, it can lie as to the meaning of the thing, being by nature *tendentious*, never as to its existence"; as such, "every photograph is a certificate of presence."[25] At the same time as a photograph certifies presence, though, it always also *presents an absence*: it always also certifies that something that was once present in front of the camera is now absent, even though it continues to be present in the photograph itself. As Barthes puts it, "In Photography, I can never deny that *the thing has been there*. There is a superimposition here: of reality and of the past."[26]

The conceptual and phenomenological problem is that while the thing "has-been" somewhere else at some other time, in the photograph they are always just *there* in the picture, not moving at all. As Barthes puts it, "When we define the Photograph as a motionless image, this does not mean only that the figures it represents do not move; it means that they do not *emerge*, do not *leave*: they are anesthetized and fastened down, like butterflies," in a display case.[27] In the case of shrine portraits, where we know we are looking at a portrait of someone alive in the picture who is now dead, the connection to butterflies in a display case is more than metaphorical. Like the pinned-down butterfly, the person in the pictures will *live on* in the photographs as a spectral presence, but the person will never again either *emerge* from the picture or *live in front of a camera* again.

Gillian Rose argues that photographs "bring near those far away," but only need to do this because their referents *are* far away in either time or space. Thus, while everyday photographs "are in some ways very full of what they show, they also produce an effect of emptiness" when the person encountering them knows that "the person *there in the photograph* is *not there* . . . (spatially or temporally) at the moment of viewing."[28] John Berger calls this awareness of the "abyss" between the moment of making and viewing a photograph the "shock of discontinuity."[29] The wider the gap in space and time, the greater the awareness of the discontinuity between the time/space of production and the time/space of viewing. The greater the discontinuity, the greater the need to bridge it, but also the stronger the feeling of ontological dissonance—that shock of discontinuity—as it is bridged.

Few photographs and viewing practices can bridge such an abyss, but sometimes it happens, which creates its own shock. This is what Barthes calls the *punctum*.[30] In *Camera Lucida*, Barthes identifies two different elements possibly contained by photographs: the *studium* and the *punctum*. The *studium* carries the culturally structured representational meanings contained in the image, while the *punctum* punctures these representational meaning structures and processes to interrupt a distanced/distancing mode of interpretation with a moment of intense feeling that feels beyond representation, whose "effect is certain but unlocatable" (and is thus difficult to discuss using words and established cultural codes, which are the tools of the *studium*).

Barthes says that "a photograph's *punctum* is that accident which pricks me (but also bruises me, is poignant to me)," so it makes sense to connect the *punctum* to trauma.[31] The *punctum* is like a portal into the sensed but unseen Real. As Barthes puts it, "The *punctum*, then, is a kind of subtle *beyond*—as if the image launched desire beyond what it permits us to see."[32] The result is an encounter with what Barthes calls the "blind field."[33] The "blind field" is exactly what the *punctum* evokes: that heterotopic and heterochronic other space of affect that is not "there" in the image but nonetheless is animated in and by the image. Barthes says that the experience of the *punctum* performs a kind of "animation" that breathes life into an otherwise lifeless representation. As he is theorizing it here, animation is a mutually constituting process similar to the process of transference: the photograph is neither animated itself nor made animated. The animation is constituted in the encounter between viewer and photographic object, located within neither the object nor the person looking: "The photograph itself is in no way animated (I do not believe in 'lifelike' photographs), but it animates me: this is what creates every adventure with photography. Whether or not it is triggered, it is an addition: it is what I add to the photograph and *what is nonetheless already there*."[34] This situation, where something is there and not there, and where time and space melt together in confusing ways, is complexly paradoxical, which is perhaps why Barthes says that the *punctum* is a "sibling to madness."[35]

Any photograph is a potential site for a mutually constitutive animating experience of the *punctum*, but because the *punctum* is not something located in a photograph itself, it only emerges within the very particular performative space of encountering the photograph. Nonetheless, the *punctum* is much more likely to emerge in encounters with photographs of people we know are dead: the person is dead, and in this picture they were alive, and will die—it's all there implicit in the photograph to begin with and only fully realized once the portrait of a past present becomes the eternal future of "being dead."

In engaging Barthes' concept of the *punctum* to theorize cultural hauntings, Avery Gordon writes that photographs of missing persons use the indexicality of photography to represent a reality that poses a question: "They have been here once. They should be here now. Where are they?"[36] Gordon argues that the "blind field" marks an absence of something present elsewhere: the blind field "is precisely what is pressing in from the other side of the fullness of the image displayed within the frame; the *puntcum* only ever evokes it and the necessity of finding it," but cannot directly *show* it.[37] The blind field is made manifest as a material present reminder of absence, a haunting, "in the paradoxical experience of seeing what appears to be not there."[38] As soon as you *know* that you are looking at a picture of someone now-dead/then-alive, the material proof of the catastrophe of photography arrives on the scene.

And that is exactly what road trauma shrines do to *every* photographic portrait of victims: shrines materially show that the person in the photograph *was then* alive when the picture was taken, but *is now* dead. In a roadside shrine, the ontological shift is there on the roadside for all to see, making it appear materially self-evident: the person pictured here is dead, and they died right here.

Here, I want to slow down and tread very carefully, because it gets to the heart of my larger argument about the melancholic affect present at road trauma shrines. At first, the recognition of the "catastrophe" of photography material-ized in shrine portraits leads to an appreciation of the first-level poignancy of a shrine photograph: here is a photograph of a person who believed in the future when this picture was taken, but because it is in a shrine, I know that future does not exist for them in any embodied form. But the next step is even more poignant. The catastrophe is not only that this person is dead but also that the shrine evidences the fact that someone is actively working-through the trauma of that loss in the present. However, the most melancholic dimension of this encounter with the *punctum* is that although the catastrophes of death and mourning are materially apparent, the depth of feeling and specific feel-ings of loss are neither represented nor transferrable to the viewer *in* or *by* the photograph. The *punctum* is a force field animated in and by photographs, not a thing contained in photographs. Ultimately, as I will explore much more fully in chapter 7, what makes shrines melancholic is that they try so hard to convey something to strangers that they can't possibly convey through objects alone.

"BEFORE" PICTURES

The fact that photographs are used to conjure absent people and objects for people located in different times/spaces is critical for how they work in road trauma shrines. Consider again the photograph we started the book with, that picture of the boy and his braces. There is nothing traumatic *in* that picture, but the picture nonetheless is all about trauma. How can that be? When photographic portraits like this are incorporated in road trauma shrines, they materialize trauma by juxtaposing *what was* with *what is no longer*. They picture a world *before trauma*—the bodily trauma that killed the mourned person *and* the trauma that the mourners left behind are medi-ating through the shrine.

Pictured within the frame of the photograph, then, is *a world with a present and projected future that does not (yet) know trauma*, which is the source of their vicarious trauma for secondary witnesses encountering them *after knowing of the trauma*. Facing a victim portrait in a shrine, the fact of the trauma is apparent even when its specific character is not: here in this photograph, this person faces a future I as a witness live within but they do not, and the main factor not only separating our realities but also connecting them is an intervening trauma that is not pictured here but is nonetheless everywhere present at the site. In short, portraits included in shrines *reflect, refract*, and *perform* trauma, but they do not *picture* trauma.

Indeed, this distinction is crucial for understanding the kinds of photographs that are incorporated into shrines and the ones that are not. Let me start with the most obvious, and therefore most deeply culturally rooted, naturalized boundary. Although shrine photographs are thoroughly entangled in the process of trauma and mourning, they do not themselves directly *represent* trauma or mourning. Situated within a road trauma shrine, photographic portraits of crash victims convey trauma without representing it. They mobilize presence through absence, by using the indexicality of

Figure 4.13. Left: U.S. Highway 285, North of Pojoaque, NM, USA, March 2006. Right: North Loop Boulevard @ Avenue F, Austin, TX, USA, February 2008.

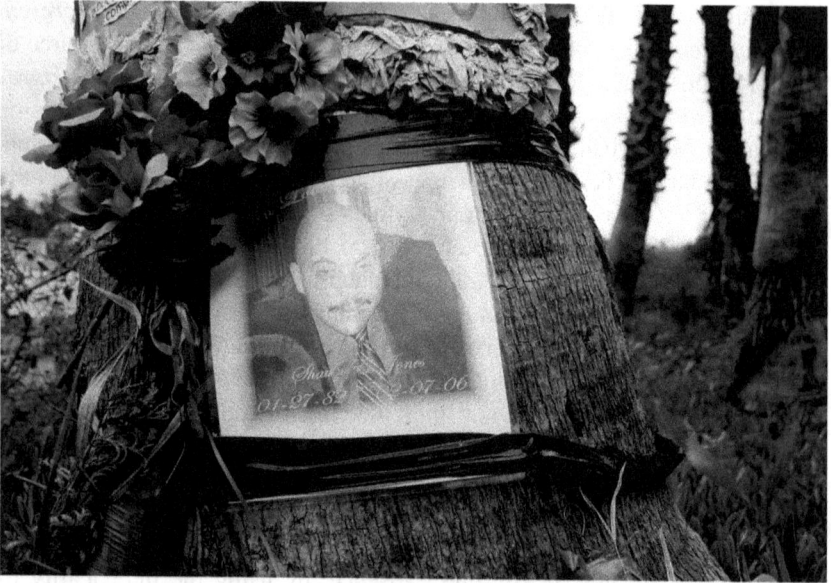

Figure 4.14. Interstate Highway 5-South @ California State Highway 163, San Diego, CA, USA, July 2006.

photography to show that what once was, is now not; what once was *becoming*, is now *done*.

This is something that photographs in road trauma shrines share with photographs mobilized in other trauma shrines and memorial museums. In *Memorial Museums* (2007), Paul Williams distinguishes two kinds of photographs used in contemporary practices of commemorating and staging traumatic violence: "action photographs" and "identification photographs."[39] Action photographs depict the processes of trauma as they occurred—violent displacements, injuries, killings, bodies that have been visibly traumatized. These pictures usually are taken by photojournalists or amateur bystanders who directly witness the trauma: Matthew Brady's photographs of Civil War carnage, Robert Capa's image of a falling Republican soldier in the Spanish Civil War, U.S. Army photographers' pictures of Buchenwald survivors, and video and mobile phone pictures of police violence at traffic stops. Sometimes these kind of action shots are part of an effort by perpetrators to document their work, as in the case of the Abu Ghraib torture pictures.[40]

Identification photographs, on the other hand, are usually produced *before they are known to be significant*, in circumstances far removed from the traumatic violence that makes them significant only in retrospect. Beginning their lives as bureaucratic headshots, family portraits, snapshots, or now even selfies, they only emerge *after* a traumatic event, where they are used to either locate or memorialize missing people. Prominent examples here would be the photographs of the missing dead from 9/11 on missing posters and memorials, the family photographs of murdered Jews in Holocaust museums, and portraits of disappeared women in the Borderlands.[41] Williams says that by "starkly displaying the faces of victims, the hope is that a visual reminder of their humanity will encourage viewers to appreciate the gravity of what occurred."[42] Williams argues that when identification photographs are seen in these new contexts, they themselves are usually read as tragic, but the tragedy is not located within the frames of the photograph: "Tragedy is more of an emotional after-effect based on the power of what we know to have happened."[43] Such photographs are chosen for their aptness in visually conveying a person's everyday identity, which then becomes the locus of their affect once their referents are absent. However, by definition, such a photograph cannot contain any actual referent to the intervening violent trauma that has made the photograph now appear tragic.

At the hundreds of roadside shrines I have encountered, I have never seen a picture at a shrine that shows the aftermath of a crash, something like a police or coroner photograph—a photograph that would *represent, display,* and *be* trauma at the same time. They are always identification pictures, never action pictures. The portraits *materialize* the fact that trauma and mourning are occurring, and would not exist at all if not for the trauma they mourn, but the pictures at shrines are pictures of life and not pictures of death. This is true of all portraits in all roadside memorials—not only in vernacular roadside shrines but also in official state memorials that include photographic portraits, such as police memorials. They all are pictures that evidence trauma only because they show a world *before trauma.* They are pictures that use the rhetorical power of photography to certify the person as once existing in the past to activate feeling about their loss in the present. They are "Before" pictures without an "After."

The photographs incorporated into road trauma shrines are ordinary, everyday photographs that would otherwise not be present in a public space such as the roadway. As such, they carry with them the discursive form of ordinary photographs even as they are also recontextualized in particular ways on the roadside. This is something they share with other trauma memorial forms in contemporary culture, especially in a media environment dominated by social media. In *Doing Family Photography: The Domestic, the Public, and the Politics of Sentiment* (2010), Gillian Rose traces the social practices of "doing" family photography as everyday photographs move from domestic, familial spaces to the mediated public sphere. Rose argues that family photographs, or "family snaps" in the British colloquial, are "entering public spaces of display more and more often. Once mostly restricted to being looked at only by the family and friends of the people pictured, family snaps are now visible more

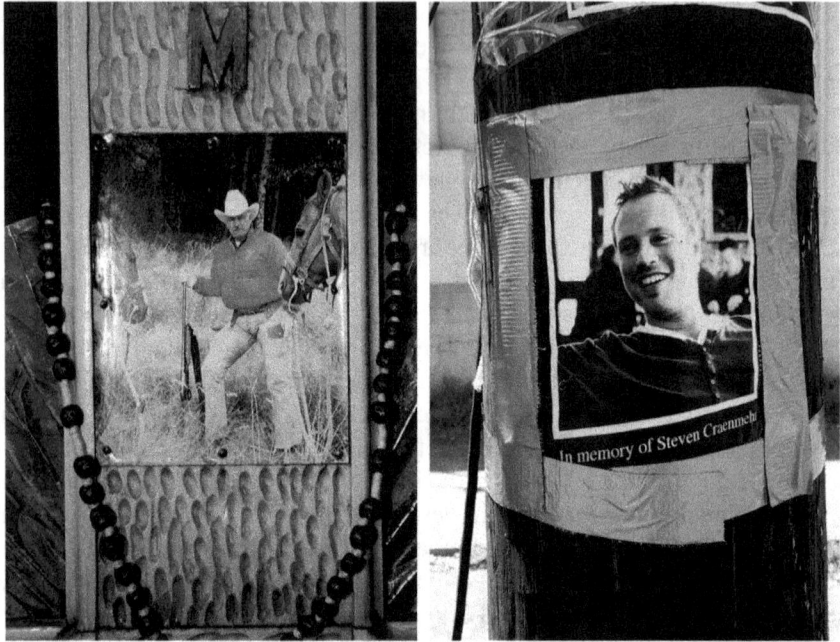

Figure 4.15. Left: Old Santa Fe Trail Road, Santa Fe, NM, USA, July 2006. Right: *SXSW 2014 DWI Crash Victim Steven Craenmehr*, **Red River Street @ E. 10th Street, Austin, TX, USA, June 2014.**

and more often to the gaze of strangers."[44] As the "visual economy" of family photographs has stretched to circulate these photographs through physical and virtual global and local places and media, the social practice of "doing family photography" has also stretched. This has implications for not only how people "do" family photography but also what is done to family photographs as material and virtual objects, including shrine portraits.

Moreover, what makes shrine portraits "identification" pictures is not only that they show the person being themselves in ordinary pictures but also that they demand witnessing by *inviting identification* by witnesses. Shrine portraits invariably show people being alive, being who they are, living their lives, showing themselves to the camera, demanding that the viewer recognize them as unique individuals. They almost always show the person smiling, as if they are demanding a (virtual) relationship with the viewer. As such, they are what Gunther Kress and Theo van Leeuwen call "Demand pictures": pictures that interpellate viewers directly, where the person is represented as simultaneously having, desiring, and demanding a relationship with YOU who look at the picture.[45] They make this demand even when in this case it is not literally true—when the demand is simply a virtual relationship, which you can accept or deny much differently than you could if the person were physically standing right there in front of you—but the demand is there nonetheless.

Seeing the subject of the picture as well as the picture itself as an object and a subject in any photographic encounter is the key to understanding how photographs can seem both inanimate and animate at the same time. In each of the photographs of photographs in this section, the subjects look into the camera, posing and knowingly showing themselves to the camera to be seen and photographed by or for the people who know them well, for the people who will see these pictures in the future and *feel something* based on their relationship to the persons pictured in them. Being photographs, they can show more (and less) than the subject and the photographer intend to show, but at their core, they show a conscious act of self-representation recorded in a conscious act of photographing, and they show a conscious attempt by both to reach beyond the frames of the photograph to project an imagined relationship with viewers who will necessarily be located in some other time and place when they encounter the photograph.

For me, the shock of discontinuity is located in their eyes. Like photographs, eyes are objects and means at the same time: they are things that we use to see and do things. Moreover, eyes allow us to see not only what is there but also beyond what is there. Looking at shrine portraits even as a stranger, being one of those viewers in the present looking at their eyes looking at you from the past, you can feel their eyes looking at you, mirroring your look at them. Alfred Gell argues that pictures representing eye contact have a complex relationship to agency and subject/object relations: "Eye-contact, mutual looking, is a basic mechanism for intersubjectivity because to look into another's eyes is not just to see the other but to see the other seeing you (I see you see me see you see me etc.)." Even represented eye contact, Gell writes, "seems to give direct access to other minds because the subject sees herself as an object, from the point of view of the other as a subject."[46]

At the same time, however, the material structure and spatial location of the shrines remind us that any "demand" the people in these photographs appear to make upon us will always now be a frustrated, projected demand. If the subject of the photograph had outlived the photograph instead of the other way around, we might one day encounter the person in the world and look them in the eye. That fact is projected within the photograph, but not realized. Placed in a road trauma shrine, a photographic portrait shows a different story, a story that frames these processes in such a way that they look at least ironic and, more often, poignant: the only reason we are even encountering this picture of a person making a virtual demand is that the life of the person in the picture has been subsequently negated, leaving them present only in a photograph, looking into a future that they will never see themselves, and looking out of a picture that they themselves will never see again either. While when posing they might have visualized themselves seeing themselves in this picture in some other place and other time, that will no longer happen for them. The eyes looking into their represented eyes will never again be their own.

It is the paradox at the core of shrine portraits: to claim the eternal life promised by the reality effects of photography, the shrine must also show that the photograph has outlived the subject of the photograph. Elizabeth Hallam and Jenny Hockey argue that this is true for other contemporary death memorial practices as well: "Memory practices can be regarded as attempts to counter loss caused by death, making connections with absent individuals and bringing them into the present; but in doing so they simultaneously evoke the gaps they have left behind."[47] But where other memorial practices may exist within the institutional structures that make the material claim of absence structurally and spatially apparent (i.e., the boundary markers of a cemetery, the page layout of an obituary page in the newspaper), the task of every roadside shrine is to not only *be* a shrine but also to both *materialize* and *show* that it is a shrine, and to do so almost always on the roadside, in the public right-of-way. The claim for eternal life "frozen" in the photograph is dependent on the shrine showing that the shrine marks the place where the person died. That is, before it can open up a space for bridging the gap between the living and the dead, the photograph must first show, *materially*, that the bridge is necessary.

Every photographic pose projects a future time and space of viewing, but cannot possibly predict or control those sites and times as a photograph "carries its referent into the uncharted future," as Ulrich Baer puts it.[48] A subject who looks out at us from a photograph demonstrates that they are conscious that they are being photographed, and thus they present themselves to be photographed, but they cannot know what will happen to their photographic likeness once the shutter has snapped. And once we know that the person appearing to make this appeal in the photograph is dead, and therefore capable of nothing but a virtual appeal, the appeal itself is the location of great poignancy, because we know that they have not fared well, and that our belated response—whether cruel or kind—comes too late to matter to them.

RELATIONS OF PROXIMITY

Now that we have established the three main types of objects materialized in road trauma shrines, I want to analyze the relations between and among objects at some particularly eclectic shrine sites. We have already seen in chapter 3 how all road trauma shrines set up relations of proximity between shrine sites and the contiguous landscape. All road trauma shrines also set up relations of proximity *within* shrine sites as well; the process of producing place is also the process of "placing" things with other things.

The sites we will engage here each contain an extraordinary collection of signifying objects: figurines of the Buddha next to ET, plastic Superman cups next to Virgin de Guadalupe icons, Christmas balloons next to teddy bears, and handmade carvings and handwritten notes and family snapshots next to all of them. Like shrines themselves, the objects within shrines appear to be individualized, spontaneously self-generated, and idiosyncratic at one scale but highly patterned at another—patterned both in terms of what kinds of objects appear and how they are arranged in "relations of proximity" that bring objects from a number of different cultural locations and compose them into a simultaneous juxtaposition.

The proximal relations of different aspects of a person's identity are even more pronounced at sites that contain multiple photographic por- traits of the victim. The shrine at the opening of this book is an example of that, where the hockey portrait and the graduation portrait not only represent different stages in the person's life but also *do* different things in the composition of the shrine. Like other material objects, printed photographs take up space, are moved through space, and do both over time. This is something that has only recently been analyzed as robustly as the representational function of photography, particularly in the work of Elizabeth Edwards and her collaborators.[49] When photographs land in a roadside shrine, especially alongside other photographic objects where images are repeated, recontextualized into collages, and take on various material and presentational forms, they open up a space for repetition with a difference, which calls attention to the collision of knowledges and

Figure 4.16. Left: Pacific Coast Highway @ Sunset Boulevard, Malibu, CA, USA, July 2006. Right: South Congress Avenue @ St. Elmo Road, Austin, TX, USA, November 2012.

realities contained in the juxtaposition of multiple photographs as well as the juxtaposition of photos with other objects.

I visualize each of the shrine portraits I include in this chapter being first displayed in the intimate public surroundings of a person who knew the victim well, someone who would have a graduation picture or a sports team portrait or even a hunting picture up on the wall or in a photo album or in a desktop slideshow or on Facebook or Instagram. There, the photograph would be situated in the everyday material and virtual world where the person used to live, where the pictures would be both woven into and indexical of the discourse of what Gillian Rose calls "togetherness."[50] As Rose argues, "'Togetherness' is not just pictured by the family photo image; it is also enacted by family members as various things are done with the images"; in short, family photographs not only *picture* togetherness but also *perform* togetherness: embedded in social practices, family photographs are the representation *and* agent of group integration, both the engine of and the location for relations of proximity.[51]

Like arrangements of family photographs in an album, on a domestic wall, or on a Facebook page or Instagram feed, which put otherwise separate images "together," the photographs in shrines are not only the agents and outcome of the labor of mourning, which is its own social practice, but also always materially *next to* other things—physically juxtaposed with other photographs and things that are placed within a certain integrated composition that asserts integration as it materializes it. In shrines that also represent togetherness by incorporating photographs of multiple people within the frames of photographs, it takes a bit of work for a stranger to identify the actual crash victim or victims among the many bodies pictured in the shrine.

Again, however, this collage aesthetic is the dominant logic for entire shrines, not just photo collages. The collision of sacred and mundane, pop culture and folk culture, is particularly jarring at sites in Northern New Mexico, the place where road trauma shrines were born and where they are the most elaborated as a cultural form, and also a place long known for its vibrant folk art traditions. The effect is arresting at this site south of Taos that not only has a jolting set of juxtapositions of objects from different cultural spaces but also features a tall rock cairn built from rocks gathered nearby and debris from the crash (see figure 4.17).

In front of the cairn is the most personalized and unique handmade object: an elaborately carved wooden cross with the victim's name on it. Of course, while this particular cross is unique in itself, it is also produced in the local style, so it looks like many other crosses displayed in the area in churches, cemeteries, yards, and art shops. The cross is draped with several rosaries, which again gives the cross an anthropomorphic quality. But at the base of the cross is an odd assembly: a set of SpongeBob Squarepants bubbles next to a

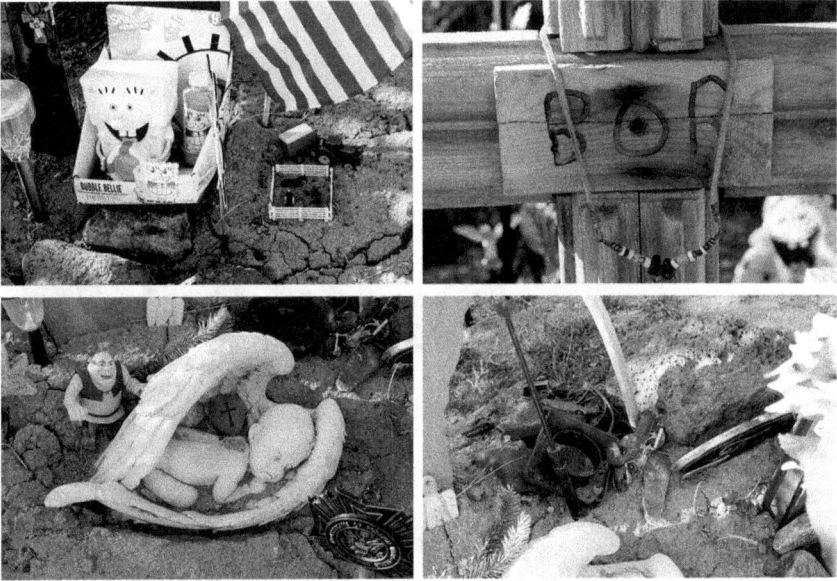

Figure 4.17. New Mexico State Highway 68-South, Pilar, NM, USA, February 2012.

carefully staged miniature farm scene. On the other side of the cross, a Shrek doll stands over a cement sculpture of a baby sleeping like a fetus in the arms of an angel. I can't decide if Shrek looks menacing here or protective. And then behind it all is another assemblage: a collection of cracked black plastic left over from the crash, topped with a blob of melted aluminum next to a plastic motorcycle figure. To the side of that gathering is the Ford nameplate from the crashed vehicle. As such, the site seems a perfect embodiment of Foucault's heterotopia.

Relations of proximity sometimes undercut or confuse the affective intensity of individual objects, but more often they amplify and deepen affect. That is certainly the case at the following site in South Texas, where several objects are attached to two trees that are themselves obviously violently scarred from the crash. In among the Latin Catholic folk art is a shocking pairing of a Bible and a child's sock. Taken alone, each object has its own powerful affect. The Bible is nailed to the tree, draped with a rosary. The edges of the pages are curled and the nails are rusted; it looks fragile and powerful at the same time. The sock is crumpled. It just slouches there, lifeless. But the proximity of the two objects ensures that the two objects are "read" together. When they are, it becomes clear that the Bible is open to Psalm 22—the Psalm that begins: "My God, My God, why have You forsaken Me?"—and the two objects buzz together with intensity.

Figure 4.18. U.S. Highway 77A-North, South of Goliad, TX, USA, May 2013.

I would like to end this chapter with a single photograph of a juxta-position of objects at a shrine in central New Mexico that has become an icon for my thinking about the way materialization works in road trauma shrines, especially relations of proximity within these shrines (see figure 4.19).

Figure 4.19. U.S. Highway 84-West, North of Santa Rosa, NM, USA, July 2006.

In the foreground is what appears to be the main focus of the picture: a whimsical ceramic and metal angel, stretching her arms to present a little blue bird, but also seeming to be inviting a hug. On the left side of the frame are plastic flowers and two Polly Pocket dolls encased in plastic bags. These figures are ringed with silk flowers. Behind the angel and to the right of her in the frame are several broken and charred pieces of the car involved in the crash: a brakelight lens cover, gears from the transmission, a brake pad. As I look at these crash relics, my eye always starts in the middle of the picture, but always stops abruptly in the bottom right corner. There, lit by the sun, is a bent car key still stuck into the ignition switch. Both are heavily charred, with plastic melted in chunks around the metal.

Seeing the ignition and key melted together sets me off each time I look at this photograph. I flash to a hand reaching in to turn the ignition for the last time. The person attached to the hand wouldn't be thinking that, of course. I think of the nonchalant way the driver would have turned it, the same way I do every day. I visualize that act, as thoughtless, automatic, and unproblematic as any ordinary action that we do in contemporary American culture—like flipping a light switch or tying your shoes—and project out a fateful drive that ended with the ignition and key forever wedded together in a material representation of the end of the line. The key, that unique identity object that not only allows the driver to drive their lives but also serves as a gatekeeper, the one that allowed the journey to commence at exactly the time and place it did, set off a chain of events that moved the family inexorably to their death there on the side of the New Mexican road. I imagine an old horror or thriller movie, where someone fleeing a killer turns the key and nothing happens, only a rrruuuu, rrruuuuu, rrruuuu, until then presto, engage, and the car springs to life. I wonder what would have happened if the driver had been delayed even a few seconds more, altering the chain of events just enough.

But I know this is a fantasy, a projection in time and space that actually takes me *away* from the trauma in the photograph, and takes me away from the trauma embedded in the materiality of the things pictured in the photograph. When I return to the photograph, and return to the materiality of the site, I can see that the memorial objects that come to rest temporarily at a shrine site are structurally analogous to the vehicles and bodies that originally collided at the crash site, where multiple mobilities converged from different trajectories into a simultaneous copresence of moving objects.[52] There, as in all road trauma shrines, an amazing array of objects is assembled to create a complex use of space, and a complex juxtaposition of objects. It creates a heterotopic space where there's a fascinating play of visual/material relations of proximity among transference objects—a place where gears and melted keys temporarily occupy the same physical and cultural space as angels and Polly Pockets.

Road trauma shrines are places where worlds collide in the same way that vehicles sometimes do on the road, but also the way the culture collides inside of each of us today. As Doreen Massey argues, "If space is a product of practices, trajectories, interrelations, if we make space through interactions at all levels, from the (so-called) local to the (so-called) global, then those spatial identities such as places, regions, nations, and the local and the global, must be forged in this relational way too, as internally complex, essentially unboundable in any absolute sense, and inevitably historically changing."[53] This is certainly apparent today, as we navigate the global flows of capital, markets, products, objects, media, and bodies, but it is important to say that the spaces of automobility have always been a contact zone, the one place where people from all levels of society exist together in a momentary and nervous tension that sometimes breaks down. The goal in analyzing road trauma shrines is the same that is there in analyzing any phenomena in automobility: to see not only the complexity of intersections, confluences, convergences, displacements, and discord of our present moment and place but also its continuities with the past and elsewhere as well as its disjunctions.

Road trauma shrines like the ones we have analyzed in the chapter not only *contain* and *consist of* an enormous array of signifying objects from different scalar systems, but they also communicate whatever they do *with* and *through* these material objects. If this and other shrines are materialized eulogies, they are collage or dialogic eulogies, not monologues. They are exceptionally multiplicitous even for the collage form—with different relationships encoding different identities and using different symbolic resources to communicate different things about the same person at different times. The objects are juxtaposed against each other in intriguing ways to create sites that are most directly only held together (as in cohering into a "text') by their particular location in space and time. And, as we will see much more clearly in the next chapter, that "holding together" itself is something always in process—whether because of the effects of climate and weather the site is eroded, or moved; or because people who maintain the sites change them periodically; or because different people, with different relationships to the person, decorate the site differently or remove things from it.

In this respect, it is important to recognize that the identities represented at shrines are always already *displaced* identities. People who build the sites speak to and on behalf of the dead, but the dead do not speak for themselves. Thus, the relations of proximity at the shrines are always already refracted through the identities of the people who "do" the memorializing and what they imagine, believe, assert, and remember—as well as what they forget.

NOTES

1 See Daniel Miller, *Stuff* (Malden, MA: Polity Press, 2009).

2 Michel Foucault, "Of Other Spaces," *Diacritics* 16 (Spring 1986): 22–27.

3 Kathleen Stewart, *Ordinary Affects* (Durham, NC: Duke University Press, 2007), 87.

4 Kathleen Stewart, "Worlding Refrains," in *The Affect Theory Reader*, ed. Melissa Gregg and Gregory J. Seigworth (Durham, NC: Duke University Press, 2010), 343.

5 Ben Highmore, *Ordinary Lives: Studies in the Everyday* (London: Routledge, 2010), 119.

6 Ben Highmore, "A Sideboard Manifesto: Design Culture in an Artificial World," in *The Design Culture Reader*, ed. Ben Highmore (London: Routledge, 2009), 2.

7 Peter Redman, "Affect Revisited: Transference-Countertransference and the Unconscious Dimensions of Affective, Felt, and Emotional Experience," *Subjectivity* 26 (2009): 61–62.

8 Ibid., 63.

9 Gillian Rose and Divya Tolia-Kelly, "Visuality/Materiality: Introducing a Manifesto for Practice," in *Visuality/Materiality: Images, Objects and Practices*, ed. Gillian Rose and Divya Tolia-Kelly (Farnham, 2012), 3.

10 Foucault, "Of Other Spaces."

11 See especially Katarina Damjanov, "Lunar Cemetery: Global Heterotopia and the Biopolitics of Death," *Leonardo* 46, no. 2 (2013): 159–62; Kelvin Knight,

"Placeless Places: Resolving the Paradox of Foucault's Heterotopia," *Textual Practice* 31, no. 1 (2017): 141–58; Nicholas Paliewicz and Marouf Hasian, "Popular Memory at Ground Zero: A Heterotopology of the National September 11 Memorial and Museum," *Popular Communication* 15, no. 1 (2017): 19–36; Joseph Pugliese, "Crisis Heterotopias and Border Zones of the Dead," *Continuum: Journal of Media & Cultural Studies* 23, no. 5 (2009): 663–79; Brent Allen Saindon, "A Doubled Heterotopia: Shifting Spatial and Visual Symbolism in the Jewish Museum Berlin's Development," *Quarterly Journal of Speech* 98, no. 1 (2012): 4–48; and Elizabethada Wright, "Rhetorical Spaces in Memorial Spaces: The Cemetery as Rhetorical Memory Place/Space," *Rhetoric Society Quarterly*, 35, no. 4 (2005): 51–81. For less spatial, more metaphorical uses of the heterotopia concept, see Ryan H. Blum, "Anxious Latitudes: Heterotopias, Subduction Zones, and the Historical-Spatial Configurations within *Dead Man*," *Critical Studies in Media Communication* 27, no. 1 (2010): 55–66; Katarina Damjanov, "Lunar Cemetery: Global Heterotopia and the Biopolitics of Death," *Leonardo* 46, no. 2 (2013): 159–62; Maria Mendel, "Heterotopias of Homelessness: Citizenship on the Margins," *Studies in Philosophy & Education* 30, no. 2 (2011): 155–68; and Gary P. Radford, Marie L. Radford, and Jessica Lingel, "The Library as Heterotopia: Michel Foucault and the Experience of Library Space," *Journal of Documentation*, 71, no. 4 (2015): 733–51.

12 See Gilles Deleuze and Feliz Guattari, *A Thousand Plateaus: Capitalism and Schizophrenia*, trans. Brian Massumi (Minneapolis: University of Minnesota Press, 1987), 474–500.

13 Jane Bennett, *Vibrant Matter: A Political Ecology of Things* (Durham, NC: Duke University Press, 2010), 24.

14 See Manuel DeLanda, *A New Philosophy of Society: Assemblage Theory and Social Complexity* (New York: Continuum, 2006), 10–11; and Manuel DeLanda, *Assemblage Theory* (Edinburgh: Edinburgh University Press, 2016). See also Martin Müller, "Assemblages and Actor-Networks: Re-Thinking Socio-Material Power, Politics and Space," *Geography Compass* 9, no. 1 (2015): 27–41. For an analysis of Deleuzian models of emergent public spheres that use roadside memorialization as a case study and engages DeLanda extensively, see Elaine Campbell, "Public Sphere as Assemblage: The Cultural Politics of Roadside Memorialization," *British Journal of Sociology* 64, no. 3 (2013): 526–47.

15 Catherine Ann Collins and Alexandra Opie, "When Places Have Agency: Roadside Shrines as Traumascapes," *Continuum: Journal of Media & Cultural Studies* 24, no. 1 (2010): 110.

16 Peter Johnson, "Unraveling Foucault's 'Different Spaces,'" *History of the Human Sciences* 19, no. 4 (2006): 77.

17 Ibid., 81.

18 Along the U.S.-Mexico border, it is not always easy to tell the difference between a roadside car crash shrine and the more general type of roadside devotional shrine associated with popular Latino Catholicism on both sides of the border, usually called a *capilla*. For more on this form, see James Griffith and Francisco Manzo Taylor, "Voices from Inside a Black Snake, Part II: Sonoran Roadside *Capillas*," *Journal of the Southwest* 48, no. 3 (2006): 233–59.

19   See Eric Mazur and Kate McCarthy, eds., *God in the Details: American Religion in Popular Culture* (New York: Routledge, 2010); David Morgan, *The Embodied Eye: Religious Visual Culture and the Social Life of Feeling* (Berkeley: University of California Press, 2012); David Morgan, *The Lure of Images: A History of Religion and Visual Media in America* (New York: Routledge, 2007); David Morgan, *The Sacred Gaze: Religious Visual Culture in Theory and Practice* (Berkeley: University of California Press, 2005); and David Morgan, ed., *Religion and Material Culture: The Matter of Belief* (New York: Routledge, 2009). See also Colleen McDannell, *Material Christianity: Religion and Popular Culture in America* (New Haven, CT: Yale University Press, 1995); and Timothy Beal, *Roadside Religion: In Search of the Sacred, the Strange, and the Substance of Faith* (Boston, MA: Beacon Press, 2006).

20   See Arjun Appadurai, ed., *The Social Life of Things: Commodities in Cultural Perspective* (Cambridge: Cambridge University Press, 1986), especially Igor Kopytoff, "The Cultural Biography of Things: Commoditization as Process," in *The Social Life of Things*, ed. Arjun Appadurai, 64–94. For an introduction to and critique of recontextualization as a methodology for visual/material culture studies, see Gillian Rose, *Visual Methodologies: An Introduction to Researching with Visual Materials*, 4th ed. (London: Sage, 2016), 253–87.

21   Margaret Gibson, *Objects of the Dead: Mourning and Memory in Everyday Life* (Victoria: Melbourne University Press, 2008).

22   Anyone from Texas driving by the shrine, which is about seventy-five miles down the road from the university, would know that Texas A&M and the Corps of Cadets, in particular, are well known for their fiercely loyal and proud school spirit.

23   See Jacques Derrida, *Limited, Inc.* (Evanston, IL: Northwestern University Press, 1988). See also Constantin Nakassis, "Citation and Citationality," *Signs and Society* 1, no. 1 (2013): 51–78.

24   Stewart, *Ordinary Affects*, 42.

25   Roland Barthes, *Camera Lucida: Reflections on Photography*, trans Richard Howard (New York: Hill & Wang, 1981), 87.

26   Ibid., 76.

27   Ibid., 57; emphasis in original.

28   Gillian, Rose, "Family Photographs and Domestic Spacings: A Case Study," *Transactions of the Institute of British Geographers* 28 (2003): 11.

29   John Berger and Jean Mohr, *Another Way of Telling* (New York: Vintage, 1982), 86–87.

30   Barthes, *Camera Lucida*, 51.

31   Ibid., 27.

32   Ibid., 59; emphasis in original.

33   Ibid., 57.

34   Ibid., 55; emphasis in original.

35   Ibid., 115.

36   Avery Gordon, *Ghostly Matters: Haunting and the Sociological Imagination*, 2nd ed. (Minneapolis: University of Minnesota Press, 2008), 109.

37   Ibid., 107.

38   Ibid.

39 Paul Williams, *Memorial Museums: The Global Rush to Commemorate Atrocities* (Oxford: Berg, 2007), 56–75.

40 See Joey Brooke Jakob, "What Remains of Abu Ghraib?: Digital Photography and Cultural Memory," *Visual Studies* 31, no. 1 (2016): 22–33. See also Ariella Azoulay, *Death's Showcase: The Power of Image in Contemporary Democracy* (Cambridge: MIT Press, 2001); John Berger, "Photographs of Agony," in *Selected Essays* (New York: Vintage, 2001); Errol Morris, *Believing Is Seeing: Observations on the Mysteries of Photography* (New York: Penguin, 2011); Mark Reinhardt, Holly Edwards, and Erina Duganne, eds., *Beautiful Suffering: Photography and the Traffic in Pain* (Chicago, IL: University of Chicago Press, 2007); Lisa Saltzman and Eric Rosenberg, eds., *Trauma and Visuality in Modernity* (Hanover: Dartmouth College Press, 2006); Susan Sontag, *Regarding the Pain of Others* (New York: Farrar, Straus & Giroux, 2003); and Barbie Zelizer, *About to Die: How News Images Move the Public* (Oxford: Oxford University Press, 2010).

41 See Carrie Boudreaux, "Public Memorialization and the Grievability of Victims in Cuidad Juárez," *Social Research* 83, no. 2 (2016): 391–417.

42 Williams, *Memorial Museums*, 71.

43 Ibid., 63. Williams contrasts this effect of tragedy with the effect of horror: action photographs provoke a sense of horror while identification photographs provoke a sense of tragedy.

44 Gillian Rose, *Doing Family Photography: The Domestic, the Public, and the Politics of Sentiment* (London: Ashgate, 2010), 4.

45 Gunther Kress and Theo van Leeuwen, *Reading Images: The Grammar of Visual Design*, 2nd ed. (London: Routledge, 2006), 117–18.

46 Alfred Gell, *Art and Agency: An Anthropological Theory* (Oxford: Oxford University Press, 1998), 120.

47 Elizabeth Hallam and Jenny Hockey, *Death, Memory, and Material Culture* (Oxford: Berg, 2001), 181.

48 Ulrich Baer, *Spectral Evidence: The Photography of Trauma* (Cambridge: MIT Press, 2005), 23.

49 See especially Elizabeth Edwards, "Material Beings: Objecthood and Ethnographic Photographs," *Visual Studies* 17 (2002): 67–75; Elizabeth Edwards and Janice Hart, eds., *Photographs Objects Histories: On the Materiality of Images* (London: Routledge, 2004); and Elizabeth Edwards, Chris Gosden, and Ruth Phillips, eds., *Sensible Objects: Colonialism, Museums, and Material Culture* (Oxford: Berg, 2006). See also Christopher Pinney, *Camera Indica: The Social Life of Indian Photographs* (Chicago, IL: University of Chicago Press, 1997).

50 Rose, "Family Photographs and Domestic Spacings." See Gillian Rose, "'Everyone's Cuddled Up and It Just Looks Really Nice': An Emotional Geography of Some Mums and Their Family Photos," *Social & Cultural Geography* 5, no. 4 (2004): 549–64. See also Pierre Bourdieu, *Photography: A Middle-Brow Art*, trans. Shaun Whiteside (Stanford, CA: Stanford University Press, 1991); Martha Langford, *Suspended Conversations: The Afterlife of Memory in Photographic Albums* (Montreal: McGill-Queen's University Press, 2001); Marianne Hirsch, *Family Frames:*

Photography, Narrative, and Postmemory (Cambridge: Harvard University Press, 1997); and Marianne Hirsch, ed., *The Familial Gaze* (Hanover: Dartmouth University Press, 1999).

51 Rose, *Doing Family Photography*, 41, 45.

52 See Kevin Hannam, Mimi Sheller, and John Urry, "Editorial: Mobilities, Immobilities and Moorings," *Mobilities* 1, no. 1 (2006): 1–22.

53 Doreen Massey, "Geographies of Responsibility," *Geografiska Annaler Series B: Human Geography* 86, no. 1 (2004): 5.

Chapter 5

Performing Road Trauma

So far, we have seen how road trauma shrines place and materialize traumatic affect by collecting and concentrating traumatic grief energy into a unique site that is built from many objects but that always is more than the sum of its parts. These are primarily material and spatial phenomena, so I now would like to address the spatiotemporal intersection of space *and* time at road trauma shrines to show how these sites are maintained, revised, and eventually displaced over time.

However, in keeping with the methodological focus of the book, here too we will be analyzing not the stories people tell about doing these performances over time, which are not located on the roadside, but the ways that such performances leave visual, material, and spatial traces at shrine sites themselves. That means that even though this chapter emphasizes performances of grief and trauma over time, it does so by analyzing the shrines themselves visually, materially, and spatially for *physical evidence of* these changes over time, focusing on a range of performative dimensions legible in material form at road trauma shrines. The focus here is on the way these shrines live their lives on the roadside until they eventually are allowed to die. The final chapter of the book contains an extended analysis of witnessing these changes at a particular site over time. My goal here, as it has been with the rest of the book, is to attune to multiple shrines as both objects and subjects in their own right. That is, once a shrine has been made into a place that magnetizes material objects with affect, people interact with it as if it is both a proxy for the deceased and a platform for communication, and that interaction leaves its own visual, material, and spatial traces than can be witnessed by anyone passing by, especially with return visits.

"LAST ALIVE" SITES

Before we can understand these interactive traces, however, we first must establish the distinctive performative character of the remediated crash site itself as a location for communicating both to absent victims and the larger audience driving by. Shrine builders consistently maintain that the site of the crash itself is central to the practice of building car crash shrines, but, as several scholars studying roadside memorialization have demonstrated, shrine builders tend to see the crash sites not as "death" sites but as "last alive" sites.[1] This is a critical distinction that has reverberating effects not only for the performance of trauma at roadside shrines but also analysis of them: road trauma shrines function as portals because they are places for the ongoing performance of a relationship, not places for mourners to "say goodbye to" or "lay to rest" either the victim or *their relationship to* the victim. Eventually, such sites will be abandoned, but not before they have done their work to both locate and function as the platform for the ongoing process of performing trauma using the shrine. This is where the shrine's unique location matters the most. And because that roadside location is also the basis of all other claims on the motoring public, it is the most important dynamic for understanding not only how traumatic grief is performed at shrine sites but also how they address wider publics.

For instance, in *Roadside Crosses in Contemporary Memorial Culture* (2002), Holly Everett argues that roadside memorial sites are not only places where people memorialize crash victims but are themselves "symbolically representative of on-going grief work."[2] Everett says that the "permanent liminality of roadside memorialization" is reflected in the fluid structure and content of roadside shrines over time, which "reflect an ongoing dialogue with the deceased."[3] Likewise, in their study of people who build and maintain roadside shrines, Charles Collins and Charles Rhine noticed that many who leave written messages at shrines address victims as if they are "departed" instead of "dead" or "deceased"—as displaced, disembodied, or transformed, but not "ceasing to exist."[4]

Everett quotes a mother of a teen car crash victim who maintains a shrine on the roadside more actively than the cemetery site because the accident site was "the last place that Nathan was" before going "straight to heaven."[5] She says she visits the site frequently not only to maintain the memory of her son but also to talk to her son to maintain her relationship with her son in a material way; she says the shrine serves this purpose well because "that's kind of where I felt his spirit was last."[6] Collins and Rhine also write that their respondents consider the act of building and visiting shrines a material sign of their continuing relationship with the victims. Thus, shrines are not a place for them to "get closure" or to "say goodbye" to the victim but a place to *show* the victim (and themselves, and others) materially, that they are actively mourning the victim—showing that, for them, the victim might be "gone, but not forgotten."[7]

Therefore, road trauma shrines serve not only as generalized communication platforms but also specifically as portals between the world of the living and the dead. As Jennifer Clark and Majella Franzmann argue, "The sacred space of [a roadside] memorial is often built upon a strong and explicit cosmology that incorporates a belief in this particular space as a kind of threshold between the world of the living and the world of the dead."[8] Similarly, Elizabeth Hallam and Jenny Hockey argue in *Death, Memory, and Material Culture* (2001) that material memory objects, sites, and practices are not only commemorations of loss but also "attempts to counter loss caused by death, making connections with the absent individuals and bringing them into the present."[9] Analyzing elaborately built and maintained gravesites that are similar in some respects to road trauma shrines, Hallam and Hockey argue that actively maintained memorial shrine sites work to "sustain the dead as socially living persons"; such sites provide "a means to maintain a physical proximity with the deceased—a sense of 'being with' a particular person now, rather than simply recalling what has passed."[10] As Hallam and Hockey argue, contemporary death memorials are places where "the 'living' deceased reside and receive visitors and gifts."[11]

Therefore, understanding how shrine sites function depends on seeing them as "last alive" sites where mourners not only wrestle with the reality of traumatic death but also perform continuing relationships with the dead at the site. In this regard, it would be best to characterize the accident site as the *trauma place*, and the place where the person's body now exists—in a cemetery or mausoleum, or in the form of ashes in an urn in the home or scattered across some other meaningful place—as the *resting place*. At long-standing road trauma shrines, this use of shrines as portals in the ongoing temporal production of place-bound communication between worlds performs a certain kind of indefinitely accumulating time, where the shrine takes on a life of its own through the continually unfolding transferential production and revision of place.

The practice of performing trauma through ongoing griefwork at a shrine site is well documented in studies of roadside shrines. What is not as well understood is how these performances are materialized at shrine sites in ways that strangers can witness even without knowing anything about the practices of intimate mourners.[12] The key to understanding how they work in both ways simultaneously is to think of road trauma shrines as a kind of *medium* for communication.

ROAD TRAUMA SHRINES AS MEDIA

Like other trauma shrines, road trauma shrines have emerged and spread throughout the culture rapidly. What started as a religiously structured practice situated within a particular ethnic culture rooted in a specific place (Latinx Catholics in Indo-Hispanic Northern New Mexico in the mid-twentieth century) has been taken up in U.S. culture more generally. The roots of the practice remain, especially in the American Southwest where most of my fieldwork has occurred, but in its current iteration, the road trauma shrine as a form is now a cultural technology put to work by all sorts of people across the country and throughout the world.

I use the term *cultural technology* very specifically here. A cultural technology is a shared material form situated within cultural practices that emerges to equip people with the tools necessary to live within the culture. The concept of a cultural technology is similar to Kenneth Burke's more specific concept of "equipment for living." Burke argued that literature equips the culture for living in the culture by providing it with narratives, characters, and ideas they could draw on to make sense of their own experience.[13] Media studies scholars have adapted Burke's concept to help explain how people use media texts as equipment for living as well.[14] For me, however, a cultural technology functions as equipment for living in a broader sense than literature or media texts do: a cultural technology is anything produced by the culture to enable members of the culture live within the culture, to manage other dynamics produced by the culture itself. There are some cultural technologies that are also seen more commonly as technologies, such as the telephone, the traffic light, the skyscraper, the cubicle, the automobile, Craigslist, and Facebook. But there are other cultural technologies less likely to appear to be technologies, such as the potluck, the birthday party, the movie premiere, the traffic code, the Sunday Skype session, the farmer's market, the selfie, and the hashtag. All of them are cultural productions that serve to make the culture function within a certain dynamic also produced by the culture.

New cultural technologies emerge when a society outgrows its existing dominant or residual technologies or develops a new set of practices or technologies that have reverberating effects across the culture developing around them.[15] The need for a technology often precedes its appearance but does not magically cause it to appear or determine how it will be taken up (or not). With a successfully integrated cultural technology, early adopters recognize its uses and make it into an emergent cultural technology, and as it spreads from there, it reaches "critical mass" and becomes dominant. Once dominant, it becomes normative and seems natural, as if it was always needed and always there as a tool to be used, masking the fact that it was constructed. Also, as it becomes dominant, the mutual constitution of technology and user makes the technology evolve while it also shapes and is being shaped by the surrounding cultural environment as well.

The early history of roadside shrines reveals that they were an adaptation of another mobile death ritual: the placement of crosses at the site where pallbearers rested while they moved a body from one place to another in Northern New Mexico. That is, they are an adaptation of a practice for the then emergent car culture that was also rooted in the practice of marking the way that moving bodies come to rest at certain places. Significantly, like the *descansos* practice it was adapted from, the practice of placing a shrine at the site of a car crash has always had two functions: it is placed by an insider who was involved with the bereavement process to mark the place as significant within the mourning process, which simultaneously also generalizes knowledge of the significance of the site to the larger community who later pass by the site and know what it signifies.

While it is clear that shrines are working as a cultural technology that helps individuals negotiate road traumas, they are also, at the same time, through both similar and different means, working as a cultural technology that could help the larger culture negotiate the collective, cultural loss of road trauma. Important for both functions is the fact that the car crash shrine has grown out of a vernacular tradition and been adopted by others as a vernacular form. As a death and bereavement ritual, a trauma shrine helps ordinary people grieve troubling deaths. There are other cultural technologies that do the same: cemeteries, crematoria, funeral services, eulogies, online memorials. But for whatever reason, those are not sufficient to contain the trouble that the ongoing fact of ubiquitous road trauma has created in the culture.

The practice also could have stayed localized to its origins, but that did not happen either. Instead, people from all walks of life now see the road trauma shrine as an apt choice, even now a "natural" choice, to address their needs as grieving, traumatized subjects. It is also important to notice here that shrines have not fully replaced those other cultural technologies, only augmented them. The need for road shrines may have preceded their arrival on the scene, but now that they have emerged, they already seem naturalized within the physical and cultural landscape. Indeed, the growing ubiquity of the shrines now brings a new challenge to address: a presumed knowing based on familiarity, which may make them appear to be obvious enough to obscure the need to ask what they are doing.

Moreover, road trauma shrines are a particular kind of cultural technology: they are a *medium* for communication. Like other communication media, shrines are both prosthetics and platforms. Material beings themselves, they extend the communicative capabilities of everyone who uses them in terms of both time and space, giving users a visual/material/spatial means of working-through trauma through embodied communication. Shrines are a medium for placing, materializing, and performing road trauma that serves as a contact zone for the multiple publics who use them.

At the microscale, the cultural work that shrines do as a medium is to provide grieving subjects with a recognizable cultural form to use to mediate the trauma and grief triggered by losing a loved one in a violent car crash. This is important work, but it is work that has been studied extensively by other scholars of roadside memorialization. What I am focused on here is the cultural work roadside crash shrines do on a much larger scale. At this macroscale, the cultural work that shrines do as a medium is to make private memory and mourning of road trauma something potentially public—something shared by the whole culture, not just the grieving subjects within it. Indeed, road trauma shrines are a primary cultural medium for making private road trauma public.

When a shrine serves as a cultural technology for multiple producers over time, the performance process itself leaves and erases its own traces in ways that are materially evident even to strangers, which is the focus of this chapter. Because these relational performances are most explicit in messages written to the dead, to other known mourners, or to larger publics, we will start our analysis of road trauma performances with them. Such messages bridge the processes of materialization, performance, and interpellation because they are often written on or part of offerings, relics, or portraits while interpellating a specific imagined audience.

However, after we analyze those written messages, we will attend to the more subtle traces of performance visible at road trauma shrines. Road trauma shrines, like other place-bound cultural forms, reflect local and translocal identities in their material objects *and* in the ways their processes of communication change over time. Like other forms that express, materialize, and contain identity, trauma shrines create and recreate fluid identities that circulate translocal, transpersonal, transcultural forms that are nonetheless clearly *located* in space and time, but not necessarily *bound* by them. Therefore, after I analyze roadside messages, I will analyze a set of shrine sites that I have photographed a number of times over the course of my fieldwork. These recursive photographs reveal the visual, material, and spatial evidence of the lives of these shrines as they function as locations for and material signs of the ongoing, place-bound performance of trauma at roadside shrines.

WRITING TRAUMA

The prevalence of handwritten messages at roadside shrines is the clearest sign that they serve as portals not only between the world of the living and the world of the dead but also between the different moments of the present. The most prevalent form of roadside writing is a message or group of messages written with Sharpie permanent markers on relics, offerings, or portraits at the site. Often these messages gather more messages, so some sites are covered in words. Some shrines even include a box of Sharpies on site to encourage the practice (see figure 5.1). The central cross is the most often used platform for these Sharpie messages, but they also appear on a number of different objects, including portraits in the shrine.

The structure, mode of address, and voice of shrine messages provide indirect evidence of the ways people both conceptualize and perform grief and trauma at these sites. They use the shrine as a portal to speak directly to the dead, but almost always mark an absence as they seem to assert a presence. You may recall that we already saw one example of this in the Introduction, when we discussed the message the "Mom" wrote to her son on his hockey team portrait: "Your Dad and I miss having you to enjoy." This statement paradoxically directly addresses the boy as if he is present, but also as if he is absent.

This same mode of address—second-person voice, addressed to the person commemorated by the shrine in continuous present tense—is pervasive at road trauma shrines. It is one of many signs that shrines like this exist as ongoing locations where people work through grief, mourning, and

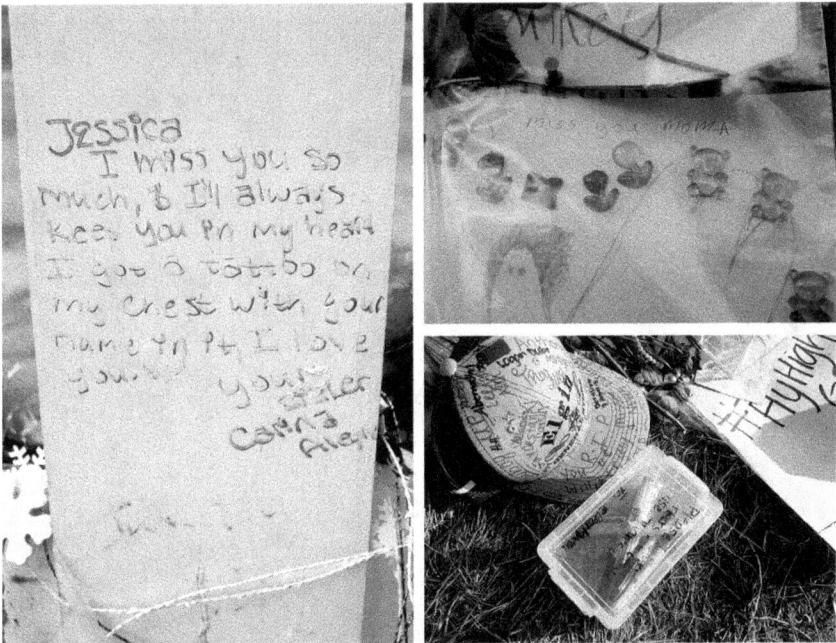

Figure 5.1. Left: E. Palmdale Road @ 90th Street-East, Palmdale, CA, USA, July 2006. Top Right: Campo Road @ Rancho Miguel Road, Jamul, CA, USA, July 2006. Bottom Right: U.S. Highway 290-East, Elgin, TX, USA, March 2015.

memory—as platforms for being with the dead at the place where they were "last alive," for communicating with them in a prosthetic form.[16] In the earlier example of the hockey portrait, there is not any explicit indication of how the Mom makes sense of the fact that the victim is simultaneously present and absent. However, many shrine messages make it clear that these writers visualize the victim having transcendent powers so that they can read these portal messages when they (or at least their spirit, ghost, soul, or angel form) "visit" the site from their new "home" in the afterlife.

That distinctive mode of address is even more elaborated in the following example from a site for a young man named Matt Huus. In the early days after the crash in late 2004, there were a range of items with Sharpied tribute messages such as, "We had some great times together. I'm gonna miss you Bro!" Among these objects was a framed handwritten message and photograph where the victim appears to be wearing the same hat that was included in the shrine right next to the picture. The message in the frame is remarkably intimate, addressing Huus as an angel that visits someone in her dreams to stay present in her life. It reads: "Hi my angel. It was so nice to see you last night. Thank you for your strength and protection. You will always be the love of my life. I miss you and already look forward to meeting you again in heaven."

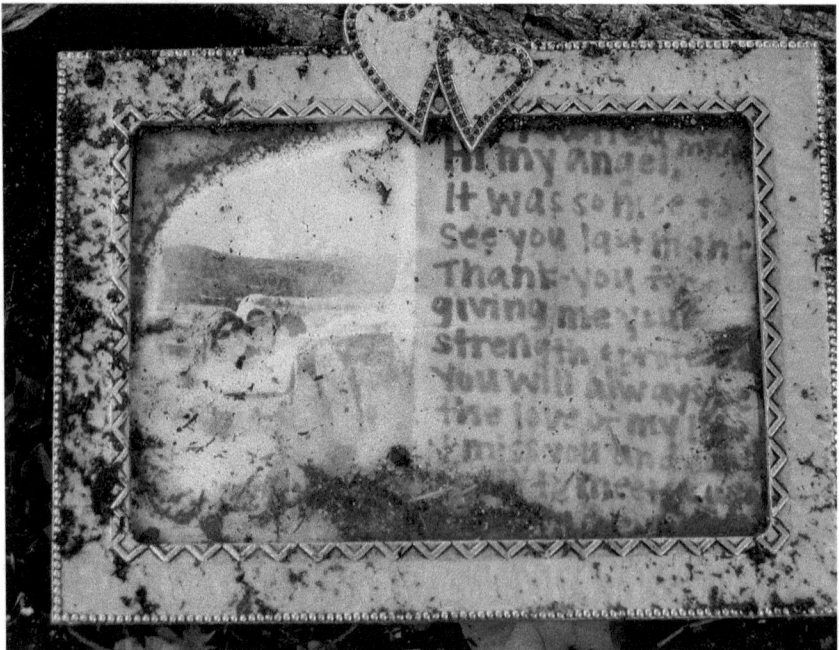

Figure 5.2. Slaughter Lane, West of Loop 1-South, Austin, TX, USA, January 2005.

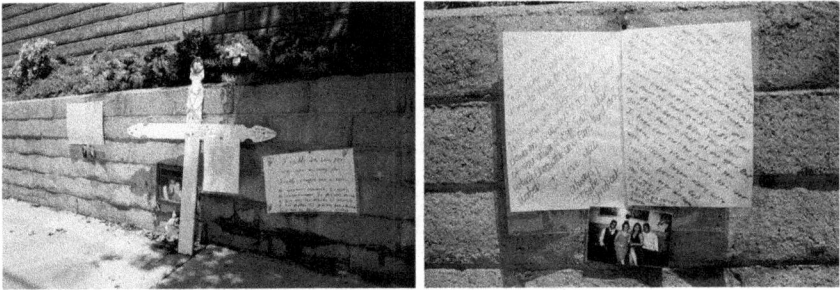

Figure 5.3. Soledad Canyon Road @ Rue Entrée, Canyon Country, CA, USA, July 2006.

Here and elsewhere, the message obviously reflects a certain religious cosmology, but sometimes it feels more like a generalized dominant American *cultural* belief: that the dead do not simply cease to exist but are transformed into different kinds of beings when they pass from this world to the next, where they can move between their world and our world at will. A shrine, like all portals, is a place where two worlds meet, and messages written at the site definitely perform a faith in the power of the portal to bridge the divides between them.

Faith in the power of the portal seems amplified when it is clear that this is not only the faith of an individual but also part of group social behavior, where mourners respond to these cues by interacting with the site alongside others or after others have already sent their messages through the portal. Sometimes these messages set up their own complex heterotopic relations of proximity, materially demonstrating relationships not only between mourners and victims but also *across* mourners.

That's what's happening in figure 5.3, where the interrelations are particularly resonant. This part of the shrine includes an assemblage of two laminated letter-size messages placed side by side above a thumb-tacked photograph of what appears to be a high school prom double date. The couple on the left of the photograph, Freddy and Lissette, died together at the site and are shown in pictures wearing the same clothes in portraits in multiple visual objects throughout the shrine, from laminated photographs to framed photographs and buttons (visible in my photograph of the whole shrine on the left in figure 5.3, and in more detail in figure 5.21).

The couple on the right of the double date picture, Isabel and Jesus, have written separate messages to their dead friends, complete with nicknames and inside jokes. Isabel addresses Lissette: "Dear Loui, I will miss you so much!!! Who is going to have my side when Jesus gets out of line? LOL!! Now you are with God looking down on us always!! I'll be thinking of you always. Especially when I am eating chocolate and can't find any water. I love you always, Isabel Chudicek." Jesus addresses Freddy: "Hey Punkass. Well what can I say. I lost a friend that was close to me like any other, but most of all I lost a brother. I mean when I had problems with my family or Isabel, the only person I could turn to was you. When I left my house I always ended up at yours and you and your parents took me in as if I was part of the family. Now that you're gone I feel a part of me gone. But I know you're not far from me. You're watching every step I take. Now you're with Lissette forever along with Jessica L. Next to God watching over all those you love!! I miss you. Te quiero un chingo, Jesus 'Chuyin' Acosta."

PERFORMING CONTINUING BONDS

While written messages are the clearest evidence of how intimates use road trauma shrines as "last alive" portals, there are other ways that the practice of "doing" shrines leaves material traces. Here I am referring to the physical evidence of changes made to shrines over time, either through intention or through wear and tear. Some shrines are continually renewed in small ways, which makes the embodied performance of trauma evident at the site even over the course of a few months. If we build on the idea that trauma shrines serve as proxies for the dead, giving them a new rematerialized life as an unfolding performance of road trauma on the roadside, shrine sites like this seem "alive and well" and "cared for," which demonstrates an elaborately staged continuing bond between mourners and the site, and thus the victim. At other sites that are not as actively maintained, tracing them over time is to watch them slowly succumb to the elements and fade away until they are entirely displaced. Other sites amount to a combination of the two, going through long periods of simply existing punctuated by fewer periodic dramatic revisions and renewals. The following site in Santa Fe, New Mexico, is like that.

At the prominent corner of two main roads on the southwestern edge of Santa Fe sits a shrine to a Harley rider named Ramon Sandoval, nicknamed "3-Wheeler Ray," who died in 1984 at the age of thirty-three. If you pause on those facts for a moment, you will realize just how extraordinary it is that this shrine still is an active site over thirty years after his death: the shrine has now lived even longer than 3-Wheeler Ray himself did. Moreover, the shrine has been in place and actively maintained long enough to become its own landmark for locals, much the way a public memorial functions not only as a place of public memory but also a taken-for-granted element of the ordinary landscape.

Since I first photographed the site in 2003, the central wooden cross has been the focal point of the shrine. Once painted red and now bleached by the bright New Mexican sun it faces for most of the day, the cross has aged considerably over the years. But the rest of the shrine continues to be renewed even as it clearly is suffering on the roadside. And the iron Harley "Ride Free" artwork to the left of the cross looks like it could be there for thirty more years or longer.

Between 2003 and 2010, the shrine didn't change much. But by 2012, the long-standing elements were redecorated and were joined by a beguiling new piece of memorial art: a striking mixed-media Virgin of Guadalupe icon mounted on a tree stump slice and fixed to a red-painted steel background. The face and hands of the Virgin were made of an unstable material, so they have shriveled into a fraction of their size, contributing to the affective intensity of the piece. On top of the cross is a pineapple woven out of palm fronds. The object itself is fascinating, but its placement gives the cross an anthropomorphic quality that resonates in a new way because of its relation to the Virgin art, particularly because there is another icon of the Virgin attached to the pineapple, which makes the pineapple look like a face, and the top of the pineapple like crazy hair.

Figure 5.4. Top: Interstate 25-North @ Cerrillos Road, Santa Fe, NM, USA, August 2003 and February 2012. Bottom: Interstate 25-North @ Cerrillos Road, Santa Fe, NM, USA, February 2012.

At other shrine sites, especially those in urban areas, the ongoing performance of traumatic grief can be made visible in a very different way. Where the site above demonstrates the continuities and discontinuities that happen within the shrine while the surrounding highway landscape stays roughly the same, at other sites the shrine seems to stay the same as the world changes around them, resolutely holding their ground as the ground shifts beneath them.

That is the case in this next shrine site in Austin, Texas, which is actually the shrine Holly Everett is referring to above when she quotes Nathan's mother talking about how his shrine works like a "last alive" site for her (see figure 5.5).[17] For almost thirty years, this site has memorialized a group of local high school students who died in a crash as they returned from lunch to the school they attended in the same central Austin neighborhood where the shrine sits today.

When I first photographed the site in 2003, the land around the shrine was still a rare large triangular greenspace in the middle of the city bounded by major streets and affectionately called "the triangle" by locals. Even here, you can see that the winds of change have come to the triangle, as the place is being prepped for a construction project. In the second photograph from a few years later, those winds have completed their work, but the shrine remains, shifted a few feet closer to the street, and now surrounded by a mixed-use development called The Triangle. It's also worth noting that the other thing that remains is the tree scar from the crash and the spray paint from the original crash investigation. The persistence—even insistence—of the shrine through these massive changes is testament to the strength of the ongoing relationship between a mother and her son, long after their relationship was transformed through a shared trauma.

The following shrine for Angelica Martinez in Pojoaque, New Mexico, is an example of sites that are continually renewed, especially in celebration

Figure 5.5. Guadalupe Street, North of 45th Street, Austin, TX, USA, October 2003 and September 2008.

of holidays, birthdays, and the anniversary of the crash. I photographed the site four separate times: March 2006, February 2010, February 2012, and July 2019, and each time it had been updated overall and was decorated to coordinate with the current holiday season. Such continual management demonstrates how a shrine can be cared for like a living family member in ways that are visible even to strangers driving by. It also demonstrates how some shrine sites set up their own little world on the roadside—an even more "different space" than most shrines—by creating a place where you can feel the presence of a site designer with a clear vision of how they intend to make a distinctive and fitting tribute to the lost victim. Shrines like this work like doll houses, train sets, or store window displays in that they are elaborately curated and staged miniature worlds where each object is carefully placed in relation to other objects to develop an overall clarity of design for the whole site. As seasons change, the design changes, but the shrine still reflects the aesthetics and identities of the people building and maintaining it. As it does, it also demonstrates, in a material way, the way that people enact their continuing bonds with the deceased and an ethical duty to care.

It is also worth noting that this particular location is in a kind of crash vortex where many shrines have come and gone while the site for Angelica Martinez has remained. Indeed, the site is actually located fifty yards north of

Figure 5.6. U.S. Highway 285, North of Pojoaque, NM, USA, March 2006, February 2010, February 2012, July 2019.

another long-standing shrine for another local Latina born the same year but killed three years later, Esperansa Sandoval. And barely visible in the background of Angelica Martinez's shrine in my photographs from 2006 and 2010 is an anti-DWI billboard featuring Polaroid-style portraits of seven other teenagers from the area who had died in recent car crashes. The billboard, sponsored by the Rio Arriba County Sheriff and the Pueblo of Pojoaque, says, "In Loving Memory of DWI Victims Everywhere. How Many Drinks is *Your* Loved One Worth?" The stretch of highway where both of these shrines are located is currently dedicated to the memory of Angelica Martinez with an official state memorial sign as well.

The site for Angelica Martinez indicates that she died on March 3, 2002. In my photographs from March 2006, there are several number 4s throughout the site, which represent the recent anniversary of the accident then. Throughout the site are also decorations for St. Patrick's Day, which was celebrated just a week before my first visit to the site. In February 2010, most of the objects in the gravel square at the base of the cross had been removed, but a giant stuffed bear and a new portrait appeared. The site was now decorated for Valentine's Day, with a new secondary structure next to the central shrine cross. The third time I photographed the site, in February 2012, the site had been significantly revised. The cross had been replaced with a new wooden cross, now minus any portrait of Angelica. The border and gravel had been raised, and new figurines appeared at the base. The site was decorated in a winter theme, with a large snowman figure on one edge of the shrine. The last time I photographed the site, in July 2019, the site was decorated for Independence Day. Figure 5.7 shows that throughout these seasonal revisions, the iconography of puppies and sunglasses provides a through line. These everyday objects that came from elsewhere always look odd out here all so neatly arranged in the median between four rushing lanes of traffic.

The following site in central New Mexico underwent a dramatic transformation in the four years between the times I photographed it. I first visited the site in the summer of 2006, a year and a half after the crash that killed four people from three different generations, presumably in the same family. Then, the site was quite large, making it visible for over a mile down the road. It included a smaller, older shrine on the fenceline, the main shrine, and evidence of the crash itself scraped into the road and still visible on the ground years later. The main shrine had been cleared of all vegetation and featured four symmetrical crosses, one for each person who died in the crash, and each cross was decorated with bursts of colorful silk flowers. At the base of one of the crosses was that assemblage of an angel, a pair of Polly Pockets, and crash debris I discussed at the end of the previous chapter. The assemblage was right in the center of the shrine, visible in figure 5.9 in a little wider view to give you context.

Figure 5.7. U.S. Highway 285, North of Pojoaque, NM, USA. Top Left: March 2006. Top Right: March 2006. Bottom Left: February 2012. Bottom Right: July 2019.

Figure 5.8. U.S. Highway 84-West, North of Santa Rosa, NM, USA, July 2006.

Figure 5.9. U.S. Highway 84-West, North of Santa Rosa, NM, USA. Top: July 2006. Bottom: December 2010.

The next time I visited the site, in December 2010, I was shaken by what I saw. The overall site had contracted significantly, and the central element—the main four-cross shrine—was gone, replaced with a single metal monument that commemorated all four victims and bricks with each person's name on them set into the ground. The crash debris was nowhere to be found, but the literal road scars remained on the road's surface, and the site continued to be cleared of vegetation. As I got closer to the site, I was astounded to find that although the debris was gone, the angel and Polly Pockets from 2006 remained, here given an even more prominent role in the more minimalist site. The combination of through lines and disjunctions is extraordinary.

There is a similar structure of continuity and discontinuity at this next site in Santa Fe, which has haunted me ever since I saw it the first year of my fieldwork, in 2003. In many ways, my continued attempt to make sense of what I was seeing at this site has led me to the focus of this chapter.

Three times since 2003—first in 2006 and later in 2010 and 2012—I returned to the site, and found it just as arresting, but almost entirely different from the site I documented in 2003. The change from 2003 to 2006 was dramatic when I first witnessed it, but the change between 2006 and 2010 was downright shocking. It was also confusing, given that the shrine appeared to have proliferated to three out of four corners of the intersection. But on closer inspection, there are also continuities, especially as certain relics and offerings continue to be present even through the massive spatial difference. The number and range of offerings and relics has decreased over time, and different elements themselves have changed, but the underlying visual/material logic and "palette" of the site have remained relatively continuous, in ways similar to the site with the angel and Polly Pockets, just over a much longer period of time. More important, because the central figure in the shrine takes on human form and not only "ages" and changes "outfits" over time, but eventually gets replaced with a more durable sculpture, the proxy function of the shrine is made literal (see figures 5.10–5.12). We will explore this process more fully in chapter 7.

CONTINUALLY LEAVING THE SITE OF TRAUMA

Of course, this process of performing road trauma through visual, material, and spatial practices does not last forever. In "Trauma and Experience," Cathy Caruth argues that the process of performing trauma produces its own trauma: "for those who undergo trauma, it is not only the moment of the

Figure 5.10. Rodeo Road @ Yucca Street, Santa Fe, NM, USA, August 2003. Northwest corner.

Figure 5.11. Rodeo Road @ Yucca Street, Santa Fe, NM, USA, February 2010. Top Left: Northeast corner, facing West. Top Right: Northwest corner. Bottom Left: Southeast corner. Bottom Right: Northeast corner, facing South.

Figure 5.12. Rodeo Road @ Yucca Street, Santa Fe, NM, USA, August 2003, March 2006, February 2010. Northwest corner.

event, but the passing out of it that is traumatic, [where] *survival itself,* in other words, *can be a crisis.*"[18] Consequently, "the trauma is a repeated suffering of the event, but it is also a continual leaving of its site."[19] Nowhere is the evidence of this process of "leaving the site" of trauma more palpable than at a trauma shrine—both while it is being actively managed and revised and when it is no longer actively maintained but still present. One day, the active griefwork will end, and the shrine will no longer work the way it once did. It will simply *be there* as a place instead of being actively revised or used as a portal. As a consequence, it will also *look* and *feel* different as well (see figure 5.13). Extending even further, if a shrine is allowed to decay into nothingness, it will be displaced entirely. Dust to dust.

In other contexts, when you see a cultural object outlive its usefulness, it usually draws its melancholic affect from that sense of being leftover, left behind, passed by. But with shrines, which materialize the fact that someone is actively performing trauma, we should look at the end of that process and the melancholy it produces in a different light. The decay of shrine objects demonstrates that the shrine is no longer needed to mediate the trauma, which can be seen as in indication that the people who have built the site have finished using the site to manage their trauma—that they have "left the site" of their trauma. Again, this is not to say that they have "moved on" to "closure" in the popular conception of this, but that they have worked through their trauma at the site enough to have learned to live with it in other ways, ways that no longer demand actively performing trauma in front of strangers in public space. This is something we will explore more fully in chapter 7. For now, I want to emphasize that the continual renewal of shrines and the continual decay of shrines have something in common: they both wear their successes and failures as proxies for lost bodies on their own material bodies as shrines.

MATERIALIZING PERFORMANCES

Living in contemporary American culture is perfectly figured in the act of driving within automobility: both are resolutely locked into the ongoing performance of an eternally unfolding present, and both are always more focused on the projected future than the leftovers from the past that move along with them, just on the other side of the "floating gap" between the present and the past.[20] While driving, the windshield we look through is over 95 percent devoted to looking into the place we are moving toward, but the rearview mirror is always there, at the corner of our eyeline, to remind us of the stuff we are leaving behind. And leftovers, in the form of relics and ruins from the past—that stuff that surfaces or resurfaces, but which is no longer useful in the same way it once was—refuse to go away, even as they slowly decay. But as leftovers surface as matter, they also somehow finally call attention to their own materiality.

This is one of Tim Edensor's main points in *Industrial Ruins* (2005).[21] Edensor writes that in a consumer culture, "the materiality of matter returns"

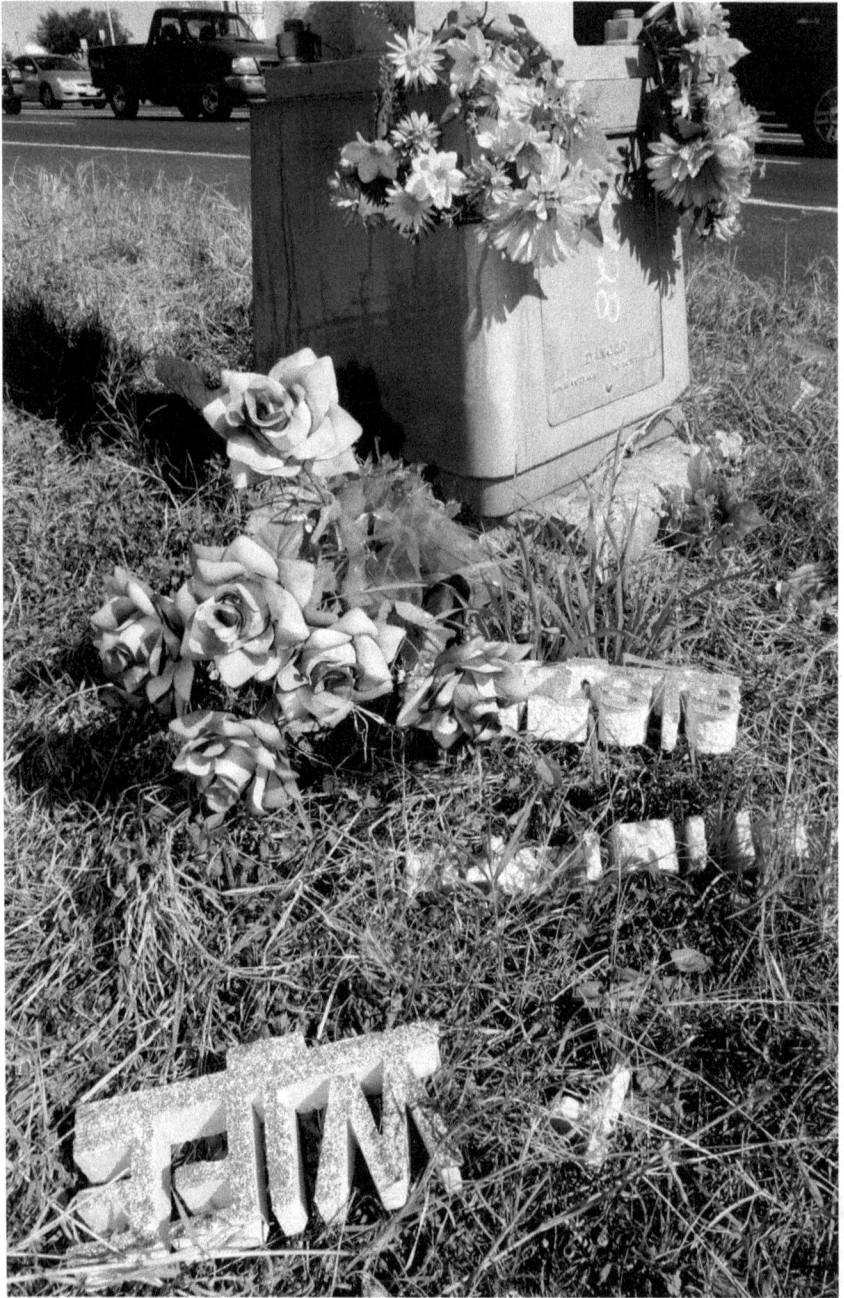

Figure 5.13. Interstate 35-South @ Martin Luther King Boulevard, Austin, USA, February 2008.

to the surface when an object is "freed from the necessity to disguise its nature in the service of commodification."[22] That is, when matter makes itself material, when either its singularized functionality is no longer taken for granted—as when a chrome exhaust pipe once mounted on a Harley is now pinned to a fence in a shrine—the fetish value of singularized objects also snaps into view. As Edensor writes, when you see "previously celebrated and valuable commodities decay and become irrelevant in the continual creation of the new, they can be recognized as the dreams they always were."[23]

The power of residual relics comes from their recalcitrance—not only from their *persistence* into the present but also their material *insistence* on being present even after "their time is up." This is particularly so at places of trauma, but it is worth recognizing that this is something they share with all physical and cultural "leftovers"—devalued detritus, refuse, and waste— as well as valued relics. Gay Hawkins argues that taking leftover objects seriously can show us "how the life of things, after we are done with them, persists and resists."[24] All of them are, as Hawkins writes, not only a form of "recalcitrant matter that refuses to go away" but also something that therefore "has an ethical claim on us."[25]

Indeed, this ethical claim is based in the fact that waste and decay are simul- taneously nouns and verbs: they are "both material stuff and process"; they are both "the things people no longer want and the inexorable force of decay and decline."[26] While the culture usually codes waste and decay negatively—both as nouns and verbs—we should also recognize that they have a kind of per- formative power, especially when they are tied to other affective objects and landscapes, as they are in road trauma shrines.

PHOTOGRAPHIC AFTERLIVES

This materialization of decay is particularly poignant with shrine portraits because they figure the lost person much more directly than any other object at a shrine. It is made even more poignant by the fact that every photograph is not only a representation but an interface and platform itself, a technologi- cally structured and materially mediated site of encounter. In *Photography: A Middle-Brow Art* (1991), Pierre Bourdieu argues that photography is a kind of social adhesive that, through interactions with particular photographic objects, "integrates" three subject/objects of photography: the person who took the photo, the subject/objects shown in the photo, and the subjects who look at the photograph as a signifying object.[27] I would add to this a fourth subject/object: *the photograph itself* as both a visual and material object, which Bourdieu makes implicit as the *site* of integration, but which to me is also an *agent* in that process. Extending Bourdieu, we could say that photographs are not only media prosthetics that "stand for" the things pictured in them as though they are proxies but also a mediating *platform* for bringing people and objects together virtually in time and space for intersubjective encounters. And extending this one step further: photographs are a mediating platform of integration for strangers as well as intimates, just in different ways.

Shrine portraits create a platform for being with the dead, for communicating with them in a situation that is not that much different from how people use photographs to bridge time and space with people who are alive, but with a crucial ontological difference: here, you know that the person shown to be alive in the picture is dead. This distinctive ontology is something we analyzed extensively in the previous chapter. Here, I want to focus more on ways that this ontology stimulates a spatiotemporally and existentially complex performative dynamic that leaves its own traces over time on the portraits that make it possible.

All photographs are used to virtually connect people by bridging gaps in both space and time, but this is particularly the case with photographs of the dead, which are even more obviously about making the absent present, about keeping the past present, and about simultaneously staging, being the location for, and acting as agents within ongoing relationships that are always more obviously virtual than relationships between living people. This is certainly how it appears to work for the "Mom" who wrote that message to her son on his hockey portrait on the California roadside: "Your Dad and I miss having you to enjoy." Before the photograph can make the absent crash victim present, it must first present that absence in the shrine. But as it does, the photograph (and the messages written on it) also shows the presence of those who do the grief work to keep the absence present in their lives as well as the lives of all who pass by the site.

Like other beings, photographic portraits at roadside shrines can simultaneously seem both powerful and fragile. They powerfully attempt to speak back to time's inevitability, but they visibly and materially succumb to it themselves out on the roadside. While I would argue that all photographs have lives and afterlives, they do not equally *show* that they do. How do photographs show that they have lives? By showing time, by showing space, and by showing the discontinuities between them all, which show that they have (been) moved through space and time. They perform and show Berger's "shock of discontinuity" by materializing the separation between the various times of viewing, the various times of exhibition/production, and the singular time of initial production. They show the moment when the photograph was originally taken (the moment when the subject consciously performed themselves to be photographed), the moment when the shrine builder installed the photograph at the site, and the many moments when someone sees that photograph as a part of the shrine. This is a process that itself carries with it all sorts of projections into imagined futures, pasts, and elsewheres. Shrine photographs show space by representing a different space of production than is materialized in exhibition and consumption. This is true for all photographs, but is especially apparent when otherwise private printed photographs are sutured into a public roadside shrine, where these spatiotemporal effects are matched by physical signs that the photograph as a physical object itself has been moved through space and has existed over time.

Where they once existed as either light imprinted on a negative or as digital data—either at rest within or circulating between the virtual spaces of a camera, smartphone, computer, or cloud-based app—photographs that get printed out to be incorporated in a shrine then take on a new material form that is both subjected to and generative of different forces. This is especially the case with contemporary photographs, which are now often printed at home on consumer printers instead of being professionally developed and printed. Once they have been printed this way, these once-digital photographs are much less chemically stable than older forms of photography, especially when they are outside exposed to the elements, as they always are in roadside shrines. There, on the side of the road, shrine photographs fade and discolor quickly, creating an uncanny mirroring of the process of moving the virtual into the material and then back into the virtual. Some portraits persist. Some disappear. Others are replaced. Others fade. Others decay. And especially when shrine portraits decay and disappear, that first transference of body to object is further materialized, revealing that the shrine portrait as proxy memory/space is not only living but also *dying*—all over again.

Consider these examples from a shrine for a young couple who died in a crash in Elgin, Texas, on February 28, 2015. I first photographed the site a week after the crash, on March 6, 2015, and I returned to photograph it on July 4, 2015. In the short space of four months, the photos printed on desktop printers decayed considerably. Even the two laminated photographs central to the shrine, which held up much better than the others because of their lamination, have faded significantly and are starting to separate and fill with condensed rainwater.

A photograph of the couple kissing in a booth of the local Dairy Queen, which was encased in a local store brand "Texas Tough" zip-lock bag to protect it from the elements, degraded even more quickly, making it even more ghostly in its content as well as form. And by late 2016, when I last photographed the shrine, all the photos were gone, along with all the other shrine objects except for the central cross.

The very thing that makes photographic portraits work in road trauma shrines is what kills them: they interpellate both an intimate public that knows the traumatic loss directly and a motoring public driving by, where these

Figure 5.14. U.S. Highway 290-East, Elgin, TX, USA, March 2015 and July 2015.

portraits are exposed to the elements. There, images that started as chemical patterns or digital code and were temporarily rendered in material form
slowly return to a virtual, spectral presence as they age and dissolve.

A crucial point here is that for a photograph to be displayed in a roadside shrine, *it must first take material form*. Before the advent of desktop
publishing and later digital photography and now smartphones and social
media, when most everyday family photographs were seen in some material form or not at all, this statement would be unremarkable, or maybe even
nonsensical. Perhaps the best way to see this is to compare roadside shrines
to the newer phenomenon of virtual memorials, such as the RIP profiles

Figure 5.15. U.S. Highway 290-East, Elgin, TX, USA, March 2015 and July 2015.

Figure 5.16. U.S. Highway 290-East, Elgin, TX, USA, July 2015 and September 2016.

common on Facebook. There, you see the same kinds of photographs you see at roadside shrines—frontal, "identification," "demand" pictures that show the person posing for a camera in some everyday context far removed from the trauma they are mediating. At virtual memorials, people also often use the same second-person, continuous present tense mode of address as they do at roadside shrines, just in the form of likes and comments instead of messages physically written on pictures and objects with Sharpies.[28]

But online, the movement between everyday image on a Facebook page to poignant memorial photograph is accomplished entirely within the same digital interface. There, photographs that start digital remain digital even as they are put to a different purpose. Moreover, an image on a screen—even one in a digital photo frame or on a home screen—appears to be a photograph but is simply a collection of pixels lit up with different colors and intensities, and it never adheres to a screen the way it does to a printed surface. Paradoxically, the same thing that makes a digital image intangible makes it permanent: if you display a digital photograph on a smartphone's home screen for a few years, it looks exactly the same, but if you print it out and display it outside even for a few months, its material form changes, often dramatically.

Once it is printed on a surface, a photographic object moves and is moved through the world, where it has a social life as a singularized photographic object with a unique history. Digital photographs stay the same as they proliferate and travel, but printed photographs almost always show evidence of their journey through space and time. That journey can evoke the blind field of Barthes' *punctum*. You might even say that unique, aged printed photographic portraits at shrines have what Walter Benjamin would say is not possible with digitally and mechanically produced photography: they have an aura.[29]

When I look at ordinary photographs in other contexts (even my own pictures of myself and my family), I find it easy to ignore how insistently they assert that the future exists—how they mask what Barthes calls their intractable anterior catastrophe. Faced with a photograph at a shrine, however, I am forced to acknowledge that this abstract potential catastrophe has actually come to pass, and that the catastrophe is being lived again by the photographs and other objects there on the roadside as prosthetics as they emerge, fade, and die.

Consider the following photographs of photographs built into the structures of roadside shrines—many of them fading, curled, and otherwise weathered, showing them to be objects in and of themselves and not just screens containing visual information. They are tactile surfaces. They are three-dimensional objects that contain two-dimensional representations. They not only "take up space" but also "show time." As the photographs stay in one place long enough to "age," they materialize time. To put it in Barthes' terms, photographs incorporated in shrines show that the "defeat of Time" also takes place *over time* in a process which itself leaves a material trace: an index of age. This is true in the home, where printed photographs exposed to light fade over time or show signs of handling, but photographs displayed outside accelerate the process. The very thing that makes them do things in the public landscape is what kills them: their exposure to elements present where people encounter them in the public right of way, where drivers are every day subject to the same forces that once converged to kill the person commemorated in the shrine.

Figure 5.17. Left: Mission Gorge Road @ Cuyamaca Street, Santee, CA, USA, July 2006. Right: Arizona State Highway 85-North, Ajo, AZ, USA, August 2006.

Included at shrines, photographs seek to renegotiate the finality of death by transposing the life of the person to the life of the photograph in space and time. It is as if the life of the person who has died is transferred onto the photographs and objects themselves: they are born, they live, they die. Some live long lives. Some outlive their usefulness and are abandoned. It is thus hard not to see them as indexes of the act of the traumatic mourning process as well. Photographs at roadside shrines have afterlives of their own, but they are hard lives. Representing moments in time, they also now show the effects of time as objects themselves. They form scars on the side of the road—scars that show that a trauma continues to be present and worked through. They die trying to keep people alive.

To further illustrate the significance of this point, let me move to W. J. T. Mitchell, who in *What Do Pictures Want?* (2005) argues that even though people in contemporary Western societies might disavow it, they have a magical understanding of the connection between pictures and the referents they represent. If anyone has any doubt of a persistent magical understanding of photography, Mitchell writes, "Ask them to take a photograph of their mother and cut out the eyes."[30] Mitchell observes that "everyone knows that a photograph of their mother is not alive, but they will still be reluctant to deface or destroy it. No modern, rational secular person thinks that pictures are to be treated like persons, but we always seem to be willing to make exceptions for special cases."[31]

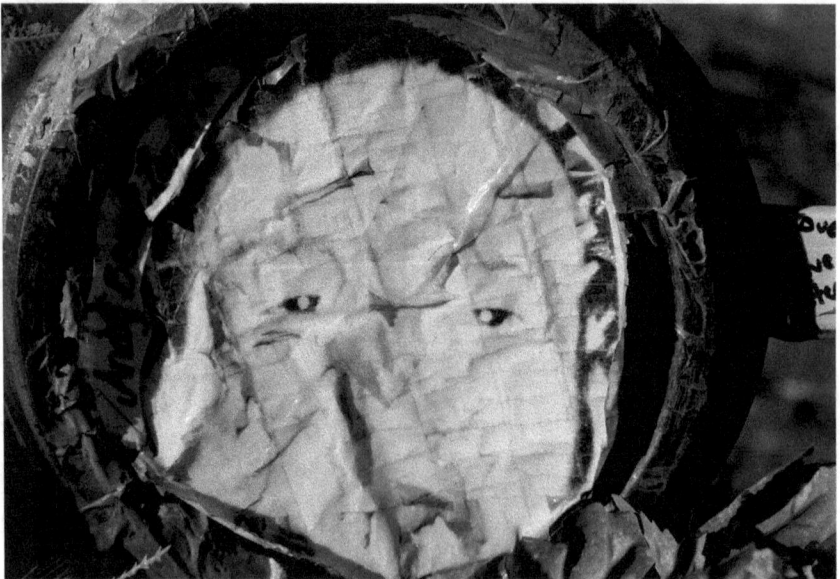

Figure 5.18. California State Highway 115-North, North of Holtville, CA, USA, July 2006.

As Mitchell's example indicates, the most recognizable "special case" category of photographs treated as "magical" is photographs of friends and family members. These are people we already know intimately as persons and not just as the subject of photographs, people we have complex feelings about—people we are already "turned toward," as affect theorist Sarah Ahmed would put it, when we encounter pictures of them.[32] And, of course, affect isn't only about affection: not only does Mitchell's example presume that we wouldn't want to harm our mothers but also ignores the many ways that his point is proven in the opposite direction, as when a former lover tears up a photograph of their ex to do symbolic violence to them.

Photographs of intimates seem to contain some of the complex affective intensity we associate with the people and objects of attachment pictured in them, giving especially photographs of dead friends and family themselves a sense of affective agency all their own. Even if we logically know that the camera cannot capture anything but visible surfaces, shapes, contours, and object relationships, we seem to see more in our own photographs than what is *in* them. Of course, we also can see *less* in photographs than others as well. As we all know from looking with blankness, confusion, or bemusement at photographs of landscapes and buildings taken by other people on vacation, the affects we feel with certain photographs are not contained in the photographs themselves, but they somehow also are still definitely activated by photographs. This is what Mitchell calls "the paradox of the image: that it is alive—but also dead; powerful—but also weak; meaningful—but also meaningless."[33]

This was true in a visual economy of images that was mostly material, such as the visual economy Mitchell presumed in 2005 when he describes cutting the eyes out of a photograph as doing both physical and figurative violence to a physical photographic object. The question more pertinent to understanding photography today, and especially the photographic portraits at contemporary road trauma shrines, is whether it is also true in the current digital media environment, where the means of producing, sharing, and viewing digital photographs have proliferated, generating an array of image forms as well as differently embodied relations with images; where images are much more likely to be produced, distributed, exhibited, and viewed virtually than materially; and where violence and affection done to virtual images is often also virtual.

In Mitchell's example, he imagines a person cutting the eyes out of a picture of their mother. It seems such a violent act, but also such a mundane act. If you turned the picture over, after all, it would become even more obvious that it is just cutting a piece of paper. But we know there is something else there. What I am proposing here is that there is yet another "something else" happening as photographic objects are materialized and age on the roadside.

Every day on the roadside, the elements are slowly destroying memorial portraits, and every day on the roadside, people who maintain shrines have to make decisions about images that are dying as they try to keep people alive.

One day, these portraits will disappear, either in a scattering wind or by the intentional hand of a grieving friend or family member who must decide what to do with these unique objects that not only materialize grief and trauma but also materialize their own history of being used to mediate a trauma. Left alone, they will fade into nothingness, giving the pictures a death that seems "natural" in its unfolding compared to the sudden shock of the victim's original death. But in most cases, there will come a day when the people who built each of these shrines must ask themselves what it means to remove a portrait from a shrine. Should it be left there, blank and mute? Should it be replaced with a new portrait?

Given the magic of photography, this is not an easy choice. It is easy in the sense that it is now very easy to make a new copy of an image that was once digital and now material, but not easy at all in the sense of determining the performativity of such a replacement. Is discarding a faded shrine portrait an act of throwing away the deceased? Is it an act of working-through, of transitioning between stages and practices of mourning, of leaving the site of trauma? Or is it an act of abandoning a continuing bond?

Figure 5.19. U.S. Highway 285-North, South of Roswell, NM, USA, August 2003.

And what of my own role as a person who is photographing these photographs at particular shrines and analyzing them—moving toward them while also being moved by them and moving them into the peculiar force field of academic scholarship to make sense of how they work, making choices about which ones "make the cut" for the book, and which ones do not? My own material and performative practices are implicated in the traffic in pictures I am analyzing. Indeed, as I will explore much more fully in the book's final chapter, my own embodied experience as a photographer has become central to my understanding of how shrines not only demand but also resist what I call dialectical witnessing.

As someone who is both a photographer and scholar, I think a lot about photography as a medium, particularly how the medium is evolving in the age of digital reproduction. Indeed, this book project has unfolded alongside both the digital revolution in photography and the explosion of social media, which have reinforced some of the medium's identifying characteristics while thoroughly transforming others. When I first went into the field in 2003, I carried two cameras with me: a Minolta SLR using 35 mm chemical film and what at the time was a fairly high-end digital snapshot camera, a fixed-lens Canon PowerShot 500. Then, the Minolta was my main camera, and I used the digital camera mainly to help me see my work in the field before I could get my photographs developed—the same way photographers and ethnographers have used Polaroid cameras in the field for years. By 2006, just three years later, I went into the field with only a Canon digital SLR for my most extensive fieldwork in the American Southwest. A Canon DSLR is still my main camera for field work, but I also now carry my iPhone with me everywhere I go now, not only for fieldwork, so I have started also using it to photograph shrines as well, especially when I happen upon a shrine in the course of my everyday activities.

While changing technologies have had implications for my own practices as a photographer, I have been even more fascinated to see how it is playing out in the *kinds* of photographic portraits I see at shrines sites. When I started doing fieldwork on shrines in 2003, shrine portraits were almost always either formal studio portraits or actual printed family snapshot photographs straight from a local film processor. Many were graduation pictures, school pictures, or team pictures, like the ones at that California site where we started the book. There, one was the rarest of all shrine portraits, a framed duplicate photo print of a professional graduation portrait. The other, the hockey team portrait, was color photocopied onto paper and laminated, which was a more prevalent form for shrine portraits already then.

Slowly, almost imperceptibly, the types of photographs and the medium on which they are printed has changed. Most shrine portraits now otherwise would exist mostly as digital images. To make them work at shrines, they are printed at home on desktop color printers. Only rarely are they photocopies or duplicates of formal studio portraits any more—unless of course the person being pictured died long enough ago to have missed the digital revolution in photography, where when are replaced, they are always a photocopy of much older formal portraits. More important, as we continue to integrate photography into our lives not only to record but also share pictures of our everyday life through social media, the photographs themselves depict a much more expansive set of scenes from everyday life. Therefore, instead of what used to be a fairly limited range of formal portraits and family photographs taken at birthdays, holidays, and vacations, there is now a remarkable range of scenes represented in these portraits and a remarkable range of photographic styles and forms, including even a form of picture prevalent now but technically impossible then—the selfie—which, because it is even more directly connected to the body of the victim as both represented and representer, is particularly poignant when it is located in a road trauma shrine.

And it is exactly this dialectic of represented and representer, subject and object, presence and absence, feeling and alienation, and knowing and not knowing that I respond to most emotionally when I photograph road trauma shrines. That's because while the people commemorated there are not there, and the people who build and maintain the shrines are not there, I am. When I encounter shrines, I also perform trauma with and through them in a different way. Like other traumas, it has taken on an obsessive, repetitive character—something I keep doing even though I know it makes me feel bad. At every site, I feel how the material reality of their presence and absence moves me, but I also physically touch the elements of the site as they touch me: I handle things, I feel their surfaces and textures, making their lives a part of my life in some small way, and I try to explore that by sometimes representing my own presence in the frame of my photographs.

Figure 5.20. Left: Arizona State Highway 86-East, Three Points, AZ, USA, August 2006. Right: Arizona State Highway 86-East, West of Tucson, AZ, USA, August 2006.

Figure 5.21. Soledad Canyon Road @ Rue Entrée, Canyon Country, CA, USA, July 2006.

Every road trauma shrine is a distinctive kind of portal, a medium—a material interface that virtually connects distant people. When I witness them, I *feel* the tangible sense of momentary copresence of two separate but connected and undeniable realities. When I am photographing them, I know that I can't picture that feeling directly, so I try to picture that "shock of discontinuity" as both a continuity and a discontinuity—an absent presence and a present absence—that applies to me as well. It is my attempt to honor the fact that these people were once there, right there where I am then standing, and that the thing that brought all of us together on the roadside was a disastrous event that transformed them from people going about their daily lives (as I am doing in the course of my research when I encounter them) to people who are commemorated in and/or perform a roadside car crash shrine.

Doing so reminds me that every time I take a photograph of a shrine, I also project a different kind of bridging of the discontinuities present at road trauma shrines: I recognize that any photograph I take of a roadside shrine projects a future for myself as an academic "doing a study" and sharing my work, as I am doing here in this book. It reminds me that, like the boy with the braces at that California shrine, I believe in the future. It is a future that imagines viewers, listeners, and readers who encounter my work in another time and place from where that work has been done. The longer this project continued, the more I began to feel that it would be a future that might not happen at all. That is because it is a future that could at any time be erased, in the blink of an eye, just as it has been for the people memorialized at the shrines I am studying. It is thus an interpellated future that could just as easily be an interrupted future.

It also is a future that would then simultaneously forever be deferred but also be read back into my photographs, where it could be experienced as a fact waiting to find a material form when all of my own photographs are located in the here and now of some other time, some other place—a time and place where I am not, but where my photographs and words *and you* are.

But I am here now, not there, then. And from here where I write, I keep coming back to the fact that what is most poignant to me about the performativity of shrine portraits is that they seem to want to convey something to strangers that they can't possibly convey through photography alone.

And, of course, I am keenly aware of this limit as I present my own photographs of photographs to you here. For me, my photographs of these shrines—especially ones where I am visible in the photograph in the form of a reflection or shadow—are often so saturated with affect for me that they are "magical." They are the sites and outcome of intense affective encounters I have had with these shrines in the field and beyond, where they were embedded in my own everyday life as a person as well as a researcher. But here in this book, where you see them now, they may appear lifeless—fastened to the page like Barthes' butterflies.

Figure 5.22. Arizona State Highway 85-North, Ajo, AZ, USA, August 2006.

NOTES

1 See Charles Collins and Charles Rhine, "Roadside Memorials," *Omega: The Journal of Death and Dying* 47, no. 3 (2003): 234; and Holly Everett, *Roadside Crosses in Contemporary Memorial Culture* (Denton: University of North Texas Press, 2002), 95–96.

2 Everett, *Roadside Crosses in Contemporary Memorial Culture*, 99.

3 Ibid., 113, 80.

4 Collins and Rhine, "Roadside Memorials."

5 Everett, *Roadside Crosses in Contemporary Memorial Culture*, 95.

6 Ibid., 96.

7 Collins and Rhine, "Roadside Memorials," 228, 235.

8 Jennifer Clark, and Majella Franzmann, "Authority from Grief: Presence and Place in the Making of Roadside Memorials," *Death Studies* 30, no. 6 (2006.): 589.

9 Elizabeth Hallam and Jenny Hockey, *Death, Memory & Material Culture* (Oxford: Berg, 2001), 181.

10 Ibid., 152.

11 Ibid., 151.

12 This work is beginning to emerge. See especially Avril Maddrell, "Living with the Deceased: Absence, Presence, and Absence-Presence," *Cultural Geographies* 20, no. 4 (2013): 501–2.

13 Kenneth Burke, *The Philosophy of Literary Form: Studies in Symbolic Action*, 3rd ed. (Berkeley: University of California Press, 1973).

14 See especially Barry Brummett, "Electric Literature as Equipment for Living," *Critical Studies in Mass Communication* 2, no. 3 (1985): 247–61; and Brian L. Ott, *The Small Screen: How Television Equips Us to Live in the Information Age* (Malden, MA: Wiley-Blackwell, 2007).

15 See Raymond Williams, *Marxism and Literature* (Oxford: Oxford University Press, 1977), 121–27; and Everett M. Rogers, *Diffusion of Innovations* (New York: Glencoe/Free Press, 1962).

16 See also Candi Cann, *Virtual Afterlives: Grieving the Dead in the Twenty-First Century* (Lexington: University of Kentucky Press, 2014), especially 81–131.

17 For more perspective on the three teenagers killed at this site and the activities and motivations of shrine builders at this and other sites in Austin, Texas, See Everett, *Roadside Crosses in Contemporary Memorial Culture*, 67–70 and 91–97.

18 Cathy Caruth, "Trauma and Experience: An Introduction," in *Trauma: Explorations in Memory*, ed. Cathy Caruth (Baltimore, MD: Johns Hopkins University Press, 1995), 9.

19 Ibid., 10.

20 The term "floating gap" comes from Jan Vansina, *Oral Tradition as History* (Madison: University of Wisconsin Press, 1985).

21 Tim Edensor, *Industrial Ruins: Spaces, Aesthetics, and Materiality* (New York: Berg, 2005).

22 Ibid., 122.

23 Ibid., 101.

24 Gay Hawkins, "History in Things: Sebald and Benjamin on Transience and Detritus," *Amsterdamer Beiträge zur Neueren Germanistik* 72, no. 1 (2009): 161.

25 Ibid., 175.

26 Ibid., 161.

27 See Pierre Bourdieu, *Photography: A Middle-Brow Art*, trans. Shaun Whiteside (Stanford, CA: Stanford University Press, 1991). See also Martha Langford, *Suspended Conversations: The Afterlife of Memory in Photographic Albums* (Montreal: McGill-Queen's University Press, 2001); Marianne Hirsch, *Family Frames: Photography, Narrative, and Postmemory* (Cambridge: Harvard University Press, 1997); and Marianne Hirsch, ed., *The Familial Gaze* (Hanover: Dartmouth University Press, 1999).

28 Cf. Avril Maddrell, "Online Memorials: The Virtual as the New Vernacular," *Bereavement Care* 31, no. 2 (2012): 46–54; and Cann, *Virtual Afterlives*, especially Chapter 4.

29 See Walter Benjamin, "The Work of Art in the Age of Mechanical Reproduction," in *Illuminations: Essays and Reflections*, ed. Hannah Arendt, trans. Harry Zohn (New York: Schocken Books, 1968), 217–51.

30 W. J. T. Mitchell, *What Do Pictures Want? The Lives and Loves of Images* (Chicago, IL: University of Chicago Press, 2005), 9.

31 Ibid., 31.

32 Sarah Ahmed, *The Cultural Politics of Emotion* (London: Routledge, 2004), 8.

33 Mitchell, *What Do Pictures Want?*, 10.

Chapter 6

Interpellating a Knowing Motoring Public

All road trauma shrines call out to witnesses because they want something. They first want to be seen, but then, like anything that wants our attention, they also want more than that. It's like the billboard for a billboard company I keep running into throughout the Southwest, the one that says, "Does Advertising Work? Just Did!" If only it were that simple. What road trauma shrines want to do is show trauma, and to show that trauma has something to do with all of us.

As I have demonstrated so far, without the indexicality of shrines as places of trauma in the landscapes of automobility, without the iconicity and indexicality of material objects at shrines, and without the material evidence of ongoing performances of trauma at shrine sites, roadside shrines would not have a means for demonstrating trauma at all. The question then becomes: to whom are they showing this trauma and how does that "showing" work?

Whether it is through the windshield of a car or on the ground at the site, road trauma shrines are witnessed where they are, as places, *in situ*, embedded within the physical as well as discursive spaces of automobility. This applies not only for intimates who encounter the site as the goal of a pilgrimage but also for strangers driving by, including researchers like me, who don't already have a preexisting reason to care about the people memorialized in the shrines when we meet them embodied in the visual/material/spatial form of a shrine. But because they are placed in the shared landscapes of automobility, roadside shrines are places where strangers visually and materially encounter a rememory—what Divya Tolia-Kelly calls an "intimate resonance" of strangers sensible through objects, pictures, and places.[1] Such a rememory conjures the stranger who died *and* the strangers who build and maintain shrines to mourn and remember them. Whatever else you know, you know that they were here, right where you drive by or where you stand—and that they died here or they constructed a shrine here. The shrine provides material evidence of both. In this way, roadside shrines, *by their very presence in the public right of way* (and even more by their form and their content), bring trauma to the rest of us by inscribing the past in the present, the sacred in the profane, the private in the public. But shrines are more than collections of visual and material objects sitting there on the roadside demonstrating trauma. They *do something*. What they do is interpellate multiple publics.

INTERPELLATION

As Louis Althusser defines it, "interpellation" is not simply a mode of addressing an audience but an *ideological* mode of address that discursively calls unique *individuals* into existence as if they are *subjects* of a projected collective.[2] While Althusser emphasizes the ways that interpellation works as a large-scale process by which individuals allow the culture's dominant ideology to represent them and collect them into a dominant ideological position as subjects so that they can be ruled, the concept of interpellation has much broader applications in contemporary scholarship in Cultural Studies and Critical/Cultural Communication Studies. As I use it in my work, interpellation refers to the process by which *any* form of communication (not just communication aligned with dominant ideology, as Althusser argues) projects an ideological space of identification and alignment. Whether it takes a linguistic, visual, or material form, interpellation addresses not individuals but *types* of people, *subjects*. To interpellate a public, therefore, is to project an ideological subject position from which the communication is *sensible* (not only in terms of capable of being perceived through the senses but also in terms of ideologically "making sense," thereby having a certain "obvious" and presumed *affective* relationship to what is being communicated). Any act of interpellation cannot determine a certain response to that positioning, but it does make a certain response seem natural. It works by "hailing" individuals as subjects within a particular discursive context and then placing those subjects within a projected collective that is positioned to not only recognize what is being communicated as appropriate and true but also to identify themselves as personally addressed by and connected to what Michael Warner calls the "indefinite address" to an imagined collective communication in some way.[3]

To interpellate a public is to call out to witnesses, and to align them with the communication being offered—placing them within a certain collective ideological and *affective* position in relation to what they are asked to witness. Like other ideologically structured forms of communication, interpellation projects a "should," and makes its projected truth seem "obvious" to anyone who witnesses it from within the projected subject position. It is performative in the same way that identities and affiliations in general both precede particular iterative performances but cannot exist without those iterative performances. That is what makes interpellation a mechanism for performativity. Recall Judith Butler's main insight into gender performativity, for instance: gender ideologies do not determine our behavior in any monolithic way, but we are called to be subjects of gender before, during, and after we perform our gender as individuals.[4] While the process of interpellation does not determine the witness's response, it does demand that the witness recognizes that they are being interpellated. If the witness "takes up" that interpellation, we say the interpellated address and the response are aligned. But the point remains that the only way that anyone would subject themselves to that projection of an "obvious should" is that they *already* share, or more appropriately *jointly inhabit*, the ideology that structures the appeal, which are never fully contained in the act of interpellation.

Interpellating subjects thus also creates a virtual affiliation *among* interpellated subjects—a group of individuals who otherwise would be strangers not only to the person, text, or object addressing them but also the rest of the strangers being addressed alongside them in the virtual public. Interpellating a public *makes a public* out of something that already potentially exists but is not yet realized as a public. As Michael Warner argues, a "public is a relation among strangers," and for these strangers to feel any affiliation with the collective that contains them alongside others they do not know, that public must be performed into being.[5] Similarly, Carole Blair, Greg Dickinson, and Brian Ott argue that the modifier "public" in public memory "situates shared memory where it is often the most salient to collectives, in constituted audiences, positioned in some kind of relationship of mutuality that implicates their common interests, investments, or destinies, with profound political implications."[6]

Sometimes these "constituted audiences" cohere into clearly marked and self-identified groups with a "naturalized feeling of belongingness" that recognize their identity as part of that audience, but the point remains that groups must always first be produced, negotiated, and maintained through specific performative acts of interpellation itself.[7] For Michael Warner, "publics do not exist apart from the discourse that addresses them" and constitutes them as a public.[8] Likewise, Gillian Rose writes that publics are only brought into existence when a mutual text, object, or place is not only produced but also mutually "articulated through a practice."[9] Therefore, as Blair, Dickinson, and Ott also argue, "A collective is posited immanently, modeled and produced by the discourses, events, objects, and practices that name and animate it."[10] Without a communicative text or object to interpellate a public, that public could not exist, but if the interpellation works, the public formed by it feels like it was inevitable.

PUBLIC AFFECTS

In addition to everything else they do, road trauma shrines interpellate publics. As they do, they also participate in a much larger trend toward performing and witnessing what might otherwise be private feelings in public physical and virtual spaces. The appeal to a public from the subjectivity of trauma and grief is a key feature of all trauma shrines, not only road trauma shrines but also many other more permanent institutionalized memorials. The most prominent U.S. example today is the group of "9/11 families," who have asserted a primary authority in shaping how Ground Zero will function as a place of public memory.[11] Such claims take shape within a larger culture characterized by publics and counterpublics formed around what Ann Cvetkovich calls "public feelings," where more and more people claim the authority to speak to, and in, the built environment and mediated public sphere from a place of grief and trauma.[12]

This is related to an even larger trend toward a kind of compulsory performance of what John Sloop and Joshua Gunn call "publicized privacy," represented most vividly in practices surrounding social media and an expansion not only of what can and should be shared in public but also even constitutes a "public" in the first place.[13] And this trend would not be happening without ubiquitous mobile communications, where the private and public interpenetrate at every level of society and we are constantly connected in virtual ways to every other person on the grid.[14] More specifically, like other suffering subjects visible in the mediated public sphere, roadside shrine builders speak to and from a culture that, as Erika Doss argues, has shifted "toward public feeling as a source of knowledge."[15] Often, trauma is claimed as what Jill Bennett calls "primary experience"—as the source of a truth claim based on an authentic voice born of trauma that has a certain authority to speak. In this, trauma truth claims are allied with other contemporary modes of public affect, where personal experience forms the basis for a nonnegotiable claim to authority.[16]

Roadside shrines speak to, from, and with a culture that operates as if these public claims of trauma are organized into economies of attention and caring that imply a scarcity model of attention and caring. It's a culture saturated with interpellation coming at us from all directions, not just from the state and from corporate media. It's a culture where we navigate our everyday lives constantly being told we should care about things—new things and entertainment commodities, for sure, but also the quotidian realities of ordinary people living their lives in the mediated public sphere: *Hey You! Look at ME! You care about me and my thing! Right?* Notice the formulation here. The framing is not a question, as is "Do you care?" but a demand for affiliation: "You are aligned with me, right?" It is also a culture where subjects compete with each other to claim not only attention but also trauma, where claims to trauma are countered with other claims to trauma in a kind of zero-sum game: my trauma is more important than your trauma, or more pointedly, our trauma is more important than their trauma.

TRACING THE MULTIPLE PUBLICS OF ROAD TRAUMA SHRINES

The salient point in relation to road trauma shrines and interpellation is that the fact that a traumatic memory or affect *exists* or *circulates* in public will not necessarily make it *public*. To make a place of public memory public, a memory place must interpellate a specific (though still indefinite) public—a collective that is made into a public in its encounter with the memory place. That is, for an individual memory to become public, it must also be performed—both *produced* and *witnessed*—in public, for a public. Witnessing individual trauma performs a shared experience, and as it does, it can produce a *collective* reality. As Frances Guerin and Roger Hallas argue in *The Image and the Witness* (2007), "The act of bearing witness is not the communication of a truth that is already known, but its actual production through this performative act. In this process, the listener becomes a witness to the witness, not only facilitating the very possibility of testimony, but also subsequently, sharing its burden. That is to say, the listener assumes responsibility to perpetuate the imperative to bear witness to the historical trauma for the sake of collective memory."[17]

Road trauma shrines, radically subjective and relational in themselves, interpellate at least two overlapping publics simultaneously. First, they interpellate a public at the microscale, where each shrine performs a private traumatic memory in public space and demands public recognition of a shared space of someone else's remembrance. The most visible collective here is the group of mourners who already knew or knew of the person before they died and who might visit and contribute to the shrine to negotiate their embodied grief at the loss of someone to whom they are already connected. However, in addition to these microprocesses at particular sites, each of these individual, localized sites acts as collective forms at a larger scale as well.

Shrines work at both the micro and macroscales to collect an already existing public that paradoxically is not yet constituted as a public, and then they create a performative space for an intersubjective encounter between the shrine and its wider public—a space for recognition of belonging to a group that suffers an individual *and* a collective trauma and loss. This second, macropublic is a very temporary, "thin," and precarious public, but a public nonetheless.[18] It is a public where those who perform and commemorate their loss demand the attention of strangers. And it is a public where strangers encounter the loss of people they will never know but nonetheless are asked to recognize as part of a collective to which they themselves belong. And, finally, it is a public made up of strangers, drivers driving by who know neither the person being commemorated nor the other drivers witnessing alongside them. That suggests that the impact (if any) of any stranger's encounter with a shrine is located not in an act of representation and recognition but in the transactive act of performing a relation between the witness and the object being witnessed or a relation across and among witnesses.

When a shrine interpellates secondary witnesses in the motoring public, it communicates implicitly through visual/material/spatial rhetoric: shrines demand we look at them and understand them, but do not do so explicitly—and certainly not linguistically. If I were to put this foundational appeal into words, it would sound something like: *Hey, You! Look at me. I have something to show you, whoever you are. Someone somebody loves died in a car crash right here. I show that someone refuses to accept it as a natural outcome of our everyday automobility. The people who built me also want you to refuse to accept it as well.* This foundational appeal becomes the platform for an additional appeal about the transaction itself, embodied in the shrine but functioning here not as a proxy of the victim, but of the grieving subjects who have built the shrine in sight of the road. This appeal might sound like this: *Now knowing trauma, I know something I do not want to know. I know something you cannot know. I know something I cannot show you. I know you will not understand, but I want you to understand. I know I can't communicate this, but I must communicate this. I want you to see what I know I cannot show you.*

In a very literal sense, people who die in car crashes do not speak for themselves. They can be witnessed but can themselves give no testimony. In this, they are like the victims of any historical trauma—those lost subjects from the past who demand witnessing in the present but can only "speak" to us through visual or material remains, through anachronistic mediations.[19] When we encounter these mediations, we are witnessing a kind of mediated testimony, making us into what Dora Apel calls "secondary witnesses" or what Andrea Liss calls "retrospective witnesses" of what Alison Landsberg calls a "prosthetic memory" of someone else's trauma.[20] Each of these concepts—secondary witnessing, retrospective witnessing, and prosthetic memory—is helpful for understanding how strangers encounter traumas they did not directly witness themselves—traumas they witness either through the mediating distance of space, through the mediating duration of time, or through the mediating effects of media technologies themselves.

To fully understand how shrines interpellate multiple publics simultaneously, we should map this out more fully in relation to car crashes and how they are witnessed. A "primary witness" to the fatality is someone who was present at the crash: someone who was involved in the physical, embodied trauma of the crash but lived through it. A primary witness who builds or contributes to a shrine would do so to negotiate their own direct trauma as they also negotiate either the trauma of losing the person who died next to them in the crash or the trauma of being responsible for killing a stranger with their vehicle (and probably also the "survivor's guilt" involved with being one who survived the crash when others did not).

All other witnesses to the traumatic loss would be considered "secondary witnesses" in relation to those primary witnesses involved in the crash (though, as we will see, there are further distinctions to be made here as well). Secondary witnesses who build and maintain shrines are directly and intimately related to the people lost in the crash but are one step removed from the bodily trauma of the crash; they thus see the trauma of the victim through the frame of at least one mirror even as the shrine evidences that they are directly negotiating a different sort of trauma: the shock of suddenly losing someone they already knew in a car crash that they themselves did not witness. The shock of knowing that someone you knew was killed so quickly and violently combined with the frustration of not knowing exactly how it happened creates its own sort of trauma. It's a cognitive trauma, not a physical trauma, but a trauma nonetheless.

For strangers who drive by, it works differently. To witness a road trauma shrine as a stranger is not to witness trauma directly, but to witness *someone else's act of performing* trauma, so witnessing is not only prosthetic but always (at least) secondary and retrospective. The same people I referred to as secondary witnesses above actually would be considered primary witnesses in relation to the strangers who witness anonymous people's trauma as they drive by. Both types of intimate witnesses—the ones directly involved in the crash and the ones who already knew the ones involved in the crash—are what Ulrich Baer calls "insider witnesses." Strangers encountering trauma shrines as they drive past them on the roadside are what Baer calls "outsider witnesses."[21] In Baer's taxonomy, an insider witness has direct testimony of the trauma they themselves have endured; an outsider witness has witnessed the trauma endured by others, either firsthand, through an embodied encounter, or prosthetically, through an encounter with a mediation (in this case the shrine).

Once insider witnesses build a shrine at the crash site, they start to generalize their testimony: whatever they think and write and do at the shrine to mediate their trauma, that performative work opens up a new space of interpellation that conveys trauma visually, materially, and spatially. And then any stranger driving by these shrines and seeing them for themselves, including me as a researcher and photographer, is an outsider witness witnessing the direct trauma of insider witnesses, but doing so in an embodied encounter. Finally, as a reader of this book, reading about and seeing photographs and reading descriptions of these particular shrine sites, you are one more step removed: an outsider witnessing an outsider witness through a mediation of a mediation.

In both examples, both the primary witnesses and the first level of secondary witnesses make up what I call the "micropublic" of insider witnesses. With car crashes, these insider witnesses bear witness to the traumatic deaths of friends and family members they were already attached to before the trauma. These are the people for whom a "Mom" flower arrangement means "My Mom," or the people who have a very specific "you" in mind when they address them in second person, present tense in written messages at the shrine. But as they do so in the public right-of-way, they also address their witnessing to the anonymous drivers who drive past shrines (the "macropublic"), the people for whom the victim will always be a stranger because they neither knew them before nor watched them experience the trauma.

At the risk of minimizing the trauma of insider witnesses, it is important to say that outsider witnessing can produce its own form of vicarious trauma. A stranger driving by a shrine can experience it as a shock, and may even be "triggered" by them through their own acts of transference. That is essentially what I felt at the California shrine for the boy in braces in 2006. But for strangers in the macropublic of outsider witnesses, the trauma of witnessing a roadside shrine is always of a decidedly different magnitude and character than the trauma of insider witnesses. Any memory they might perform through witnessing is always a prosthetic memory: a rememory at most internalized through a mediation (the road trauma shrine) that is addressed to them as if it pertains to them, but that will never fully belong to them the way a memory of the trauma belongs to the micropublic of insiders. A stranger might imagine that trauma victims are "like them" in some general humanistic sense or that they might potentially share their fate and thus "might be like them" in the sense of *momento mori*, but they will never know it is true.

Crucially, Ulrich Baer argues that public testimony can only communicate beyond insider witnesses when it is located neither only among outsider or insider witnesses but *among* both insider and outsider witnesses.[22] In other words, for a traumatic memory to be communicated and integrated into wider and wider collectives, it must bridge the two kinds of witnessing by interpellating both audiences at once, just as roadside shrines do. However, as we have already seen, with a form like a road trauma shrine, the communication is material, not representational, and structured in affect instead of (or at least in addition to) signification—particularly explicit linguistic signification. The insider witnesses will already know, firsthand, both the victim and at least some of the details of the trauma they endured. They might treat the shrine as if it signifies that directly, but any outsider looking at the shrine would know that is not true. This is the key insight W. J. T. Mitchell has when he describes the duality of photography: like a printed photograph, which can be simultaneously magic and "just a picture" (and even more radically, "just a piece of paper"), a shrine is simultaneously full and empty at the same time. What an outsider sees is that a trauma has occurred, but not the specific nature of that trauma or even the person who endured the trauma. But that is still something. And it is this "something" that forms the basis for the largest public that road trauma shrines interpellate.

HAILING DRIVERS

Intrinsic to the road trauma shrine form is that it connects not only multiple publics but also multiple *reluctant* publics. None of these collectives are made up of people who want to be in the collective. The intimate witnesses in the insider micropublic presumably did not want to lose the person who died, and did not want to perform the role of grieving "survivor." Likewise, the drivers driving by—the outsider macropublic of secondary witnesses—may not want to be "distracted" by reminders of the prevalence of road trauma as they drive down the road doing their everyday activities. As we have seen, with both kinds of publics, the main way that shrines interpellate a public is by asserting and actually materializing an ongoing social presence for the person who is "no longer with us" in body but is very much present in proxy form in the shrine.

Alberto Barrera writes that roadside shrines "are a way of saying to the departed loved one, 'I will remember you always, but I also want for the community to remember you as they come face-to-face with the cross on the roadside.'"[23] Barrera's characterization identifies the two main audiences shrine builders address: the lost person and a wider public. One is a definite address and one is an indefinite address, but in both senses, both are projected abstractions communicated with and to almost entirely implicitly and indefinitely: neither the lost person nor the wider public can be concretely addressed or known, and neither can be confirmed to have been communicated with. As we have seen, crash shrines clearly are built and used as portals where the survivors communicate with the dead, and do so, as Barrera rightly says, in both words and deeds. Second, though, shrines address a more public audience as well.

How to *locate* that wider audience is no simple matter, however. Part of the problem is our language for describing it. Barrera's use of the word "community" here implies a preestablished, localized, geographically stable collective that mourns the loss of one of its own. However, that kind of preexisting collective does not precede any encounter with a roadside shrine. On the roadside, shrines are embedded within not only public space generally but the public *roadscape*, which produces a distinctive kind of sociality, where drivers drive next to, near, and sometimes *into* strangers as they pursue their everyday lives.

In today's world of global flows of objects and people enmeshed in multiple networks of overlapping or intersecting physical and virtual collectives, we must reconceptualize the "community" that crash shrines interpellate. As we do, we must remember that the act of interpellating a public *performs that public into being* instead of addressing some preexisting and stable collective always already located in a certain time and space. This is Althusser's main point about interpellation, as well as Michael Warner's main point about publics. It is also the central feature of Judith Butler's concept of performativity. In a recent extension of her work on performativity, Judith Butler shows how loose and temporary collectives can be formed through both the act of addressing a public and being addressed as a public. Butler concedes that sometimes collectives preexist an act of interpellation, but even then they must be enacted performatively to actually function as and be felt as a collective. In many contemporary collectives, which function similarly to the *knowing motoring public* I am proposing here, "there is a collective acting without a preestablished collective subject," where "the 'we' is enacted by the assembly of bodies, plural, persisting, acting, and laying claim to a public sphere."[24] Butler observes, "If the plural subject is constituted in the course of its performative action, then it is not already constituted; whatever form it has prior to its performative exercise is not the same as the form it takes as it acts, and after it has acted."[25] That is, public claims are made to a public that is performed into being through communicative acts, so any sense of being in a collective is precarious. Finally, before such a claim can be enacted, the collective must "secure the material conditions for appearing in public" by either gaining access to a preexisting platform or creating their own.[26] In the case of road trauma shrines, that platform is the roadside itself, augmented by the now established cultural practice of building shrines to mark places of road trauma.

All testimonies that interpellate witnesses are not only accounts that represent what happened but also performative speech acts that *do something as they say something*. The term "speech act" first originated in the work of J. L. Austin in the 1960s to analyze the way that communication is both informative and performative.[27] My use of the term "speech act" here draws on that legacy while also more directly engaging De Certeau's more spatial and material use of the term to describe things like walking and renting as tactical speech acts because they enunciate an implicit, material message-through-use of space.[28] A pertinent example here would be the widely shared ritual of walking around a grave site, where to walk directly on a gravesite is seen as a physical speech act that does metaphorical violence to the dead. Thinking of speech acts more metaphorically like this as *physical communicative acts* helps explain also how material objects simultaneously represent things, are things, and do things. Every road trauma shrine performs a materialized speech act that would best be understood as what Judith Butler calls a "performative enactment": it "says" something and "does" something at the same time.[29] In some cases, the visual/material/spatial communication is amplified by a linguistic speech act, but its communication is never contained by language alone. As Butler argues, "Sometimes the bodies on the street need not speak to make their demand" because ultimately "the speech act is always doing something more and other than what it is actually saying."[30] As we will see in a moment, the linguistic messages present at road trauma shrines and other roadside memorial forms are either redundant or depend on the foundational material communication of the shrine to anchor the words.

Thinking of shrines as material communicative acts—as "performative enactments"—is also key for understanding how shrines interpellate witnesses within a relational dynamic of testimony and witnessing. As Frances Guerin and Roger Hallas write, "For a witness to perform an act of bearing witness, she must address an other, a listener who consequently functions as a witness to the original witness. The act of bearing witness thus constitutes a specific form of address to an other [that] occurs only in a framework of relationality, in which the testimonial act is itself witnessed by an other. This relationality between the survivor-witness and the listener-witness frames the act of bearing witness as a performative speech act."[31]

Whether they are based in language or in visuality, materiality, or spatiality, acts of testimony addressed to secondary witnesses do not simply *refer to* or *represent* an event. They make a materialized truth claim *as a performative act* as well. At minimum, as Guerin and Hallas write, "the performative act of bearing witness affirms the reality of the event witnessed."[32] But the act of bearing witness also "produces its 'truth' in the moment of testimonial enunciation" and the nature of that claim "depends on the discursive and institutional context in which it functions."[33] In the case of road trauma shrines, that enunciation always is embedded within automobility, where the testimony and the witness are both firmly located in the same spaces and discourses, and where any act of interpellation happens primarily through visual, material, and spatial means.

Perhaps the best way to demonstrate how all shrines implicitly interpellate multiple publics through visual/material/spatial means is to look at examples where road trauma shrines and other roadside memorials make their appeal to motorists explicitly in words: shrines that contain explicit messages about drunk driving, safety, or speed; official state drunk driving memorials; Police and State Trooper roadside memorials; Mothers Against Drunk Driving (MADD) roadside crosses; and ghost bikes. All of the examples that follow feature a much more clearly articulated mode of address, but are not as simple in their rhetoric as they might first seem. Moreover, as I will show, these elaborations and adaptations of the road trauma shrine form all depend on the already established "materially self-evident" rhetorical power of road trauma shrines in order to work. In other words, without the existence of the roadside shrine form as a recognized and useful *medium* or *platform* for showing that road trauma has occurred in a particular place, the explicit messages attached to these forms would not make sense. We will return to that point in a moment.

The most common explicit message at roadside shrines addressed to anonymous motorists is about drinking and driving (see figure 6.1). Although these appeals may seem straightforward, they're enabled by a fairly

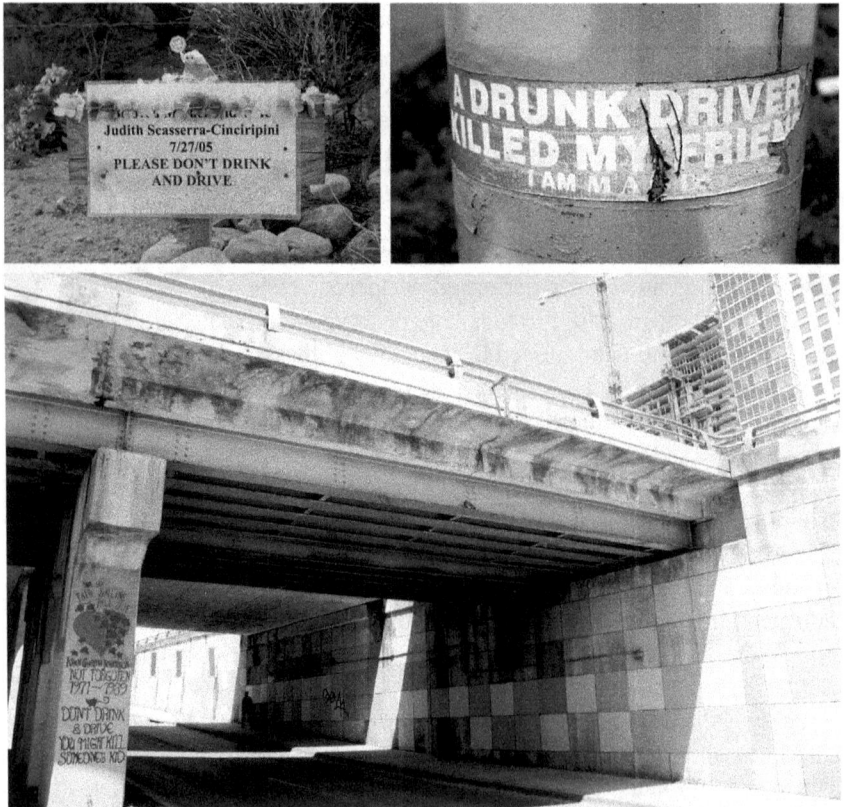

Figure 6.1. Top Left: Old Santa Fe Trail, Santa Fe, NM, USA, July 2006. Top Right: Pomerado Road @ Scripps Ranch Boulevard, San Diego, CA, USA, July 2006. Bottom: N. Lamar Boulevard @ W. 3rd Street, Austin, TX, USA, May 2015.

complicated communication process. In the top left example in figure 6.1, the "Please Don't Drink and Drive" and "Stop DWI" messages are anchored by the shrine to be more than arbitrarily placed public safety messages. In the example on the top right, a MADD bumper sticker saying "A Drunk Driver Killed My Friend. I am MADD" has been placed at a shrine.

Without the materially self-evident performative communication of all road shrines—that this is the place where someone died in a car crash, and that has something to do with you—these explicit direct addresses about drunk driving would be free-floating. This is particularly intriguing with the bumper sticker, because usually it would be seen attached to a moving vehicle as a declaration of being a grieving subject, not at the place where someone has been killed in a vehicle. By anchoring it in the shrine, the shrine is presumed to function not only as the material demonstration of authority to address an audience of strangers but also as a material demonstration of the *cost* of drunk driving. Taken together, the shrine and written messages say, *Hey, You! Look at Me! Someone Somebody Loves Died Here, and I am not Happy About It! To Avoid Experiencing My Trauma and Theirs, Please Don't Drink and Drive, and Keep Others from Drinking and Driving, too.*

This implicit spatial anchoring is even more explicit in a long-standing shrine painted on a bridge support at a prominent intersection in Austin, Texas (see figure 6.1, Bottom). This shrine, which is now one of the city's well-known downtown landmarks, was first established in 1989 and has been renewed over the years to keep it present despite the explosive growth of the city. The main message of the shrine reads, "Don't Drink & Drive. You Might Kill Someone's Kid."[34] This takes the self-evidence of the shrine form as the foundation to address drivers directly, even in this case including an explicit direct address to a "You" in the message. Here, unlike the written messages analyzed in the previous chapter, the "You" addressed is not the victim, but strangers driving by. It also makes it clear that in this case the person memorialized in the shrine was killed by a drunk driver, so it directs the message in an even more particular way: instead of "Don't drink and drive" in general, which includes those who kill themselves in drunk driving accidents, it addresses drivers as if they might one day be the *perpetrator* and would presumably like to avoid that fate. It reveals that the previous example of anti-DWI messages also presumes that you would not want to be a perpetrator, but also is vaguer than this message, meaning that heeding the message "Please Don't Drink and Drive" could keep you safe from yourself if you are the one who drives drunk, not only other drunk drivers.

Another common direct address made about public safety at road trauma shrines is "Slow Down." Like the "Don't Drink and Drive" messages, the imperative form of the message here has an implied "Hey You!" in front of it. Again, the direct message depends on the materially self-evident performative enactment of the fact that the site marks a car crash death. That is, the message is again predicated on the realization of the trauma that occurred there. Except here, it becomes even less clear how slowing down will prevent other crashes. Without knowing the particulars of the crash, it is not necessarily clear if or how slowing down would have prevented this particular crash (see figure 6.2, Left). Was the person who died in this crash, Abdias Christopher Castillo III, driving too fast when he died? Was another driver driving too fast so that they ran into him as he was innocently "slowing down?" Again, it reminds us that the "Please Don't Drink and Drive" appeal is probably always presumed to address potential perpetrators in hopes of changing their behavior, not potential victims.

Compare these more direct appeals to the invitational public safety appeal at another site, which includes a bumper sticker that reads, "I Stop on Red" (see figure 6.2, Right). The statement again appears simple but its rhetoric is actually convoluted to trace. It does not say, *Hey You! See what happens when people don't stop on red? You should stop on red!* Instead, it says, *Hey You! See what happens when people don't stop on red? I have learned that it is not OK, so I stop on red. You should join me in stopping on red, too, OK?*

All road trauma shrines perform an implicit warning to other drivers: *don't let this happen to you*. The "you" implied is multiple: (you) don't kill people, (you) don't die, and (you) don't go through the trauma of having someone you know and love die, etc. Some make these warnings linguistically explicit. The most extensive example of this I have witnessed—actually the only shrine I have ever seen that contains a detailed narrative about the crash itself at the site itself—is from a shrine I photographed in 2006 in Malibu, California. The shrine was situated on top of the granite boulders forming the seawall in between the Pacific Coast Highway and the Pacific Ocean at Malibu Beach. In the center of the shrine, next to a central cross covered in Sharpied messages, pictures, and objects, was a framed laminated 8.5 × 11-inch text ringed with pictures of the two twenty-two-year-old crash victims: Tyler Love and Keith Naylor (see figure 6.3).

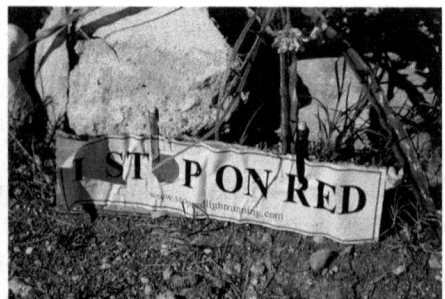

Figure 6.2. Left: U.S. Highway 290-East, Manor, TX, USA, September 2016. Right: Parmer Lane @ Texas State Highway 130, Austin, TX, USA, February 2008.

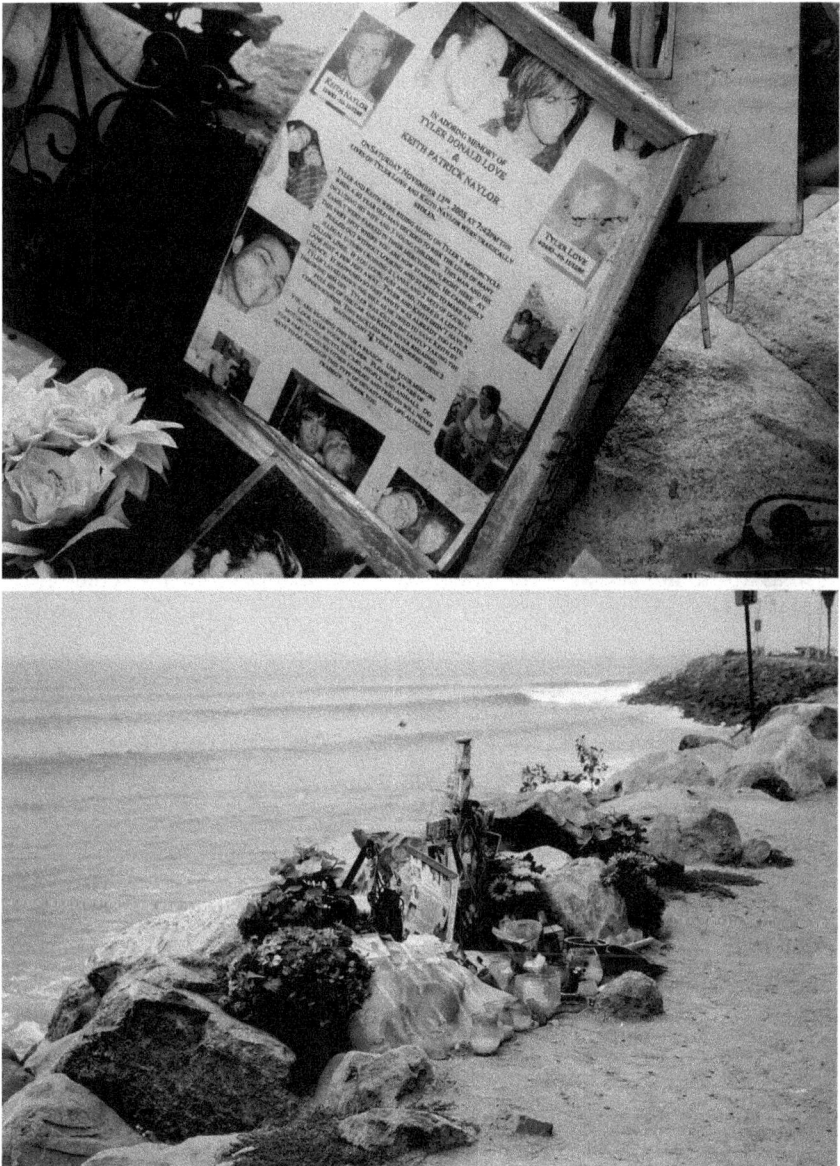

Figure 6.3. Pacific Coast Highway @ Sunset Boulevard, Malibu, CA, USA, July 2006.

Unlike the other examples of interpellation we have looked at so far, which are addressed to drivers and are actually legible from the road, the written text not only interpellates readers as drivers but also presumes that they have temporarily gotten out of their cars long enough to read the message on foot. The text starts with a narrative about the crash written by an insider witness and ends with a paragraph directly addressing outsider witnesses to the shrine. As it does, the text addresses all drivers indefinitely, but makes a negative example out of the perpetrator and sets up a politics of alignment in the interpellation, with a clear distinction between perpetrator and victims structured within an implicit generational and class conflict as well as a conflict between motorcycles and cars.

According to the author of the text, Tyler and Keith were innocently "riding along on Tyler's motorcycle," when they were "murdered" by a "reckless" forty-year-old man driving a Mercedes SUV who "decided to risk the lives of many, including his wife and 3 young children" when he "carelessly pulled out without looking," taking an "illegal U-turn" and slamming right into Tyler and Keith, who "didn't have a chance." Importantly, as the writer conveys the narrative, they address us directly in the second person as a "you" who is in a certain embodied relationship to the shrine when we are reading it. This is most evident in the way the text anchors trauma to place: The crash happened "right here. At the very spot where you are now standing." Further, to enable us to see just how reckless and careless this act was, they also ask us to notice our surroundings and imagine a different outcome if the driver had behaved differently: "If you look, just ahead, there is left turn lane just a few feet away." The narrative is followed by a final paragraph directly addressing the motoring public: "You are reading this for a reason. Use your mirrors. Look over your shoulder. Please be aware of motorcycles, bicycles, cars, people, and animals. Do your part to ensure other families and friends will never have to go through this type of devastating life altering tragedy. Thank you."

This written form of direct address to drivers is very rare. What I find most interesting about it is how it explicitly interpellates a motoring public with the goal of bringing them within a smaller subset of that public that I have been arguing is reinforced by every shrine through visual, material, and spatial means: the *knowing motoring public*. That process is happening here through pictures, objects, and relations of proximity as well while it also given linguistic form in the framed text. In many ways, the written text is redundant. Every shrine already does this communicative work without having to say it in words.

No discussion of interpellation and roadside shrines would be complete without the following example. In 2006, on my way home from my longest fieldwork trip in the Southwest, I came upon a bizarre site and sight: a spoof of a road trauma shrine on the side of Interstate 10, just east of Lordsburg, New Mexico (see figure 6.4).

Try as I might, I have never discovered who placed this intriguing mock shrine on the roadside, and I remain puzzled about its intent. My assumption is that it is a generalized *momento mori* broadcast to all drivers—an act of

Figure 6.4. Interstate 10-East, East of Lordsburg, NM, USA, August 2006.

interpellation addressing an indefinite and global "You"—but without the usual anchoring of a memorial for a lost individual. It's a warning without the mourning. In some ways, it's kin to the many Christian crosses that exist on the roadside, especially in the Southern, Midwestern, and Western United States, that are not road trauma shrines but simply roadside crosses, placed there either as a profession of faith or as an implicit challenge to accept Christianity. Read this way, the "You" would be being addressed as if they could die at any moment and should accept Christianity "before it is too late," and they head to hell instead of heaven, as one billboard nearby says. Regardless, it shows how the materiality of a roadside shrine works as the foundational platform for interpellative appeals, because it would be absolutely nonsensical if it did not cite the roadside shrine form. That is, without the implication that the mock shrine *could* represent "You," the potential victim of a car crash, the spoof would not work.

Although most states regulate the practice, road trauma shrines do a public service, which is one reason why they are allowed to stay on the roadside even though they are illegal. It's also why states have promoted their own public safety messages tied to memorializing crash victims and why states have given quasi-public status to public safety advocacy groups like MADD, which has lobbied to be exempted from most state bans on roadside memorials. There is a site near Elgin, Texas, where there is both a state memorial sign and a MADD cross for the same victim side by side, so I will use it as an example to show how both forms interpellate publics in ways similar to and different from roadside crash shrines (see figure 6.5, Left). It's worth noting here that these two quasi-public memorial forms are often used as the platform for shrine performances, making their materiality and interpellation even more complex (see figure 6.5, Right).

The blue memorial sign produced by Texas Department of Transportation (TxDOT) has a standard template that contains a public safety message about drunk driving tied to a memorial for an individual. The sign contains the following text, rendered in the same format used across the state for other official memorial signs: "PLEASE/DON'T DRINK/AND DRIVE/IN MEMORY OF/Leanne Moore Gaulden/December 13, 1999." Similar signs in other states are more oblique in their mode of address with messages like "Drive Safely"—or are tied to Highway Adoption programs, but are similar in that they mobilize the memory of a crash victim to promote public safety more generally.

The sign's interpellation is direct in some ways and indirect in others: *Hey You! I mark the spot where someone named Leanne Moore Gaulden was killed by a drunk driver. You don't know her, but you wouldn't want to drink, drive, and kill someone, would you? So Please Don't Drink and Drive or you might kill someone like a person did here.* Again, notice that the fact that

Figure 6.5. Left: U.S. Highway 290-East, East of Elgin, TX, USA, July 2010. Top Right: New Mexico State Highway 518-North, North of Las Vegas, NM, USA, December 2012. Bottom Left: Airport Boulevard @ MLK Boulevard, Austin, TX, USA, February 2010.

this is the place where it occurred is implied here, as is the fact that the date represents her death date and not some other date. Both of these implicit facts are thus entirely dependent on knowledge of the longer-standing practice of roadside shrines to connect the public safety message to the site-specific communication that happens at roadside shrines.

Unlike these official state memorial signs, which usually contain an explicit public safety message spoken in an institutional voice in an indefinite address targeted at anonymous drivers, such as "Drive Safely" or "Don't Drink and Drive," the MADD crosses are focused entirely on remembering the victim. They are both more direct in their description of the cause of the trauma they mark and more indirect in their interpellation. With their iconic white Roman crosses with red placards, MADD roadside crosses are now the most easily recognizable roadside memorial form in the American landscape. The red placard on this cross contains the following text, rendered in the same font and format used in similar crosses across the country: "IN LOVING MEMORY OF/LEANNE MOORE GAULDEN/BORN 11–5–69 & KILLED AT/THIS LOCATION 12–13–99 / BY A DRUNK DRIVER."

The explicit reference not only to death but also to *killing* also makes these crosses different from the majority of vernacular road trauma shrines, which, as we have seen, are considered by most practitioners and researchers to be "last alive" sites instead of death sites, and which, therefore, actually rarely mention death. This fact makes them already not only more like state memorial signs but also more like police investigation markings, because both focus on the cause of death more than the person who died (see chapter 3 for a fuller discussion of this). Moreover, the identification of the embodied agent of the death realizes a point only incipient in the state memorial signs: that both of these forms memorialize only victims, never perpetrators who might have died in the same crash.

The mode of address here is quite oblique, and actually hard to trace with any confidence. I realize as I render it here that almost all of it is implied: *Hey You! See me? I mark the place where a drunk driver killed a person who is missed. You wouldn't want to do that, right? So Don't Drink and Drive.* Again, the form depends on the self-evidence of the roadside shrine form to make it appear that the site is not simply either an arbitrary or convenient place to stage such a message but the site of the drunk-driving crash itself. In the end, the MADD form also focuses more on using the victim's death as a *representation* of the cost of drunk driving than on the individual themselves (which is exactly the opposite of what vernacular shrines do).

These state memorial programs and state-condoned memorial projects like the MADD crosses are matched by local and state police and trooper memorials, which use the vernacular roadside memorial form to memorialize governmental officials killed on the road in the line of duty (see figure 6.6). These memorials register a different cost of automobility—the loss of life caused in the *policing* of automobility. Unlike vernacular road trauma shrines, police memorials presume a higher value for their sacrificed life, just as war memorials do: this driver made "the ultimate sacrifice" to keep all drivers safe. Vernacular trauma shrines can reach for nothing like that; they demonstrate loss without being able to remediate it as a sacrifice to some larger social value. Also unlike private road trauma shrines, Police and Trooper Memorials are often made of durable materials like granite and steel because they are presumed to be permanent and legitimate uses of the right-of-way by organizations invested with the institutional authority to place things in the roadway with impunity.

In other ways, however, police roadside memorials are similar to the road trauma shrines they are adapted from. Most important, police memorials are located in the right-of-way, so they use the same site-specific material rhetoric to indicate that a loss has occurred *as a loss to automobility*: here, on this spot, an officer was killed doing his job to protect "us": the motoring public. As with road trauma shrines, this implicit message is materialized in the memorial's location. Including the police or sheriff's department badge on the memorial also works materially to both certify the site as legitimate and amplify the anchorage of the location to say: *here is where a law*

Figure 6.6. Left: U.S. Highway 290-East, East of Elgin, TX, USA, July 2010. Top Right: Interstate 70-East, San Rafael Swell View Rest Area, West of Green River, UT, USA, July 2006. Bottom Right: Utah State Highway 20-East @ U.S. Highway 89-South, North of Panguitch, UT, USA, July 2006.

enforcement officer was killed in the line of duty (and not simply: *this person died somewhere*). Of course, unlike most vernacular roadside shrines, most official police memorials include some kind of narrative describing how they died, further personalizing the death while still keeping it focused on public sacrifice, as you can see more clearly in the Austin Police Memorial pictured here (see figure 6.6).

The placement of police memorials in the right-of-way is what makes them function like other roadside memorials, but it is also what has led to controversies about them. The most significant example of this was the controversy surrounding a set of memorial crosses installed on the roadside by the Utah

State Troopers Association in the mid-1990s–mid-2000s. The crosses were twelve feet tall and steel, making them much larger than most private roadside memorials—more on the scale of roadside Christian devotional crosses and three-cross Calvary displays scattered across the American roadscape, which are usually also located on private land next to the roadway, not in the public right-of-way. And unlike other private roadside memorials in the right-of-way, which are located as close to the original crash site as possible, the Utah Trooper Crosses were often grouped together and/or placed at centralized locations sometimes many miles from the death sites—such as official Rest Areas, major highway interchanges, and City Parks—where they could attract more attention (and contemplation) from both drivers and pedestrians (see figure 6.6).

For all of these reasons, the Utah Trooper crosses also attracted the attention of proponents of the constitutional separation of church and state, who saw these crosses as a state-sponsored promotion of Christianity on public land.[35] The controversy was made even more interesting by the fact that it occurred in Utah, a predominantly Mormon state where the cross actually does not figure in the iconography of The Church of Latter-Day-Saints. Indeed, in an interview with me, the Troopers Association spokesman Lieutenant Lee Perry defended the practice by saying they decided to use the cross not a Christian symbol but as a generic American cultural symbol to signify that someone had died and was being remembered.[36] The Association considered the religious connotations of the cross, but decided to use the cross form anyway because they thought it signified "death" to most people in Utah, not Christianity. Their main reference point was Western movies, not religion. As Lt. Perry put it, "It's like the old Western days, when your buddy died and you tied two sticks of wood together to make an R.I.P. cross where you buried him on the side of the trail. Except now it's out on the highway. When people are driving at 65 MPH or more, what's the one symbol that shows that someone died there? A cross. And it communicates that even if drivers don't stop to see who died there . . . We thought people would see the cross and think: 'Here's where a Trooper gave his life for our safety.' "[37]

This makes it clear that the Association took for granted the material interpellative communication contained in the roadside crash shrine form, and adapted it to this more institutional speech act. After a series of court cases brought against the Troopers Association by the American Atheists, the Association eventually was forced to remove the crosses from their roadside sites on public land. The crosses ended up being relocated to a single large-scale Trooper Memorial honoring all of the fallen troopers on private land but in view of Interstate 15 in Hurricane, Utah, which is where they remain today.[38]

The final adaptation of the road trauma shrine form I would like to discuss in terms of interpellation is the phenomenon of "ghost bikes," a variety of

politically explicit "grassroots memorials" focused not only on commemorating cyclists killed in car crashes but also protesting the power differential
between cyclists and drivers within contemporary automobility (see figure 6.7).[39] The ghost bike practice started in St. Louis in 2003 and has spread
quickly from there. The ghostbikes.org website has become a clearing house
for the practice, including articulations of the strategy for the practice as well
as detailed how-to advice to help people who want to build their own ghost
bikes without having them removed by the authorities. The New York City
Street Memorial Project, which runs the ghostbikes.org website, says there

**Figure 6.7. Top Left: Kearny Villa Road @ Gun Club Road (Edge of Miramar Marine
Corps Air Station), Mira Mesa, CA, USA, July 2006. Top Right: Loop 360-North @ Bee
Cave Road, Austin, TX, USA, November 2012. Bottom Left: U.S. Highway 183-North,
South of Anderson Mill Road, Austin, TX, USA, November 2012. Bottom Right: Interstate
35-South @ Chisolm Parkway, Round Rock, TX, USA, May 2015.**

are now over 630 ghost bikes in 210 different places across the world. The bikes are most prominent in places where cyclists and motorists openly compete for mobility rights in the same scarce roadscape, such as New York City, London, San Francisco, and Austin, where the many ghost bikes there "serve as reminders of the tragedy that took place on an otherwise anonymous street corner, and as quiet statements in support of cyclists' right to safe travel" alongside automobiles.[40] Some ghost bikes stand alone as shrines attached to bikes, but some are accompanied by separate road trauma shrines, which creates particularly complex sites that interpellate multiple publics through multiple modalities.

As this description of ghost bikes makes clear, all of these adaptations of the roadside shrine form—the state memorial signs, the MADD sites, the police memorial sites, and ghost bikes—take for granted the material self-evidence of road trauma shrines, but use that implicit message and interpellation to carry an additional implicit (and sometimes explicit) messages and interpellation more specific than most road trauma shrines. For instance, like the MADD crosses, now that the ghost bike form is established enough to be recognizable as its own cultural form, the iconic white bike now serves the same anchoring function that a cross does for a more general road trauma shrine, all while interpellating motorists in a more particular way: *Hey you! Yes, You there, in your car! Look at me! Here is where someone like you in a car killed someone on a bike. You should Drive Safely, and Share the Road with Bikes!* Sometimes, this second part of the message is made literal at the sites, as when the white bike is accompanied by a "Be Kind to Cyclists" sign (see figure 6.7).

But notice again here that none of these memorial forms that adapt the roadside shrine form needs to explicitly say: "Someone Died Here at the Hands of a Driver!" That fact is performatively enacted by the roadside memorial form itself, which all of these adaptations both take for granted and cite in their extensions. That makes it clear just how naturalized the form has become in contemporary culture, particularly contemporary roadside culture.

INTERPELLATING A KNOWING MOTORING PUBLIC

A final example should solidify this point even further. State public safety messages are much more direct and explicit in their communication than road trauma shrines, and much more generally addressed to a generic or implied "You." Where a shrine shows the consequence of not driving safely by emphasizing *who* was lost in the crash, namely someone somebody loved, official messages generally use that material communication as a platform for a much more general imperative message, *(Hey, You!)(You Should) Drive Safely.* But when state public safety messages use crashes to show *why* people should drive safely, they need to use a two-part message and interpellation instead of a single materialization. That is, unlike a shrine, they must first show explicitly that a crash occurred (something a shrine does implicitly) before they can use it to try to direct drivers' behavior.

There's a powerful example of this in the small Texas town of McDade, which has seen more than its fair share of automotive fatalities in recent years. In the summer of 2016, after a series of fatal crashes at the intersection of U.S. Highway 290 and Texas FM 2236, a stoplight was installed to help prevent crashes. Even after that, a series of crashes at the intersection continued into the fall. In an extraordinary move, highway officials installed a temporary electric sign a few hundred yards east and west of the intersection on Highway 290 that cycled through two flashing messages: "Fatality Occurred Here" and "Drive Safely" (see figures 6.8 and 6.9).

We should pause on this two-part cycling message to see that it is actually a three-part message that also has three functions as a speech act: it says something, does something, and interpellates a public at the same time. The three functions are more clear if we make the three parts of its implicit mode of address explicit. The first cycle of the sign says, *Hey You! Look at Me! This is unusual and important! Now that I have your attention, Read This: A Traffic Fatality Occurred Here!* As the sign cycles, it implies a pause in the message that has the form of a rhetorical question: *So what does that have to do with you, you ask? Well, let me tell you.* When the cycle clicks over, the message is completed: *You Should Drive Safely so this doesn't happen to you, too!*

At the intersection of U.S. Highway 290 and FM 2336 itself, attached to the new stoplight, there is a shrine with the same three-part, three-function message, collapsed entirely into the visuality, materiality, and spatiality of the shrine itself (see figure 6.8).

This shrine does all the work that the cycling sign does and more. The cycling sign establishes a temporary place for communicating to drivers, says that a fatality occurred, and uses that statement as a platform to interpellate drivers with a public safety message presumed to be addressing them as a public defined in relation to the place-bound message. It does all this in an official capacity with a sign that calls attention to itself because it is temporary and thus presumably urgent, but that in other ways seems natural and strategically "proper," as de Certeau would say: after all, it is just one more example of the state communicating to drivers in a space that the state strategically controls on behalf of the public it governs, and where the state has the presumptively legitimate authority to speak to drivers often, directly, loudly, and through multiple means.

The shrine, on the other hand, makes place by tactically appropriating space it does not control, and attaching a shrine to the roadside infrastructure (place-making through building a Roadside Attachment). The shrine uses

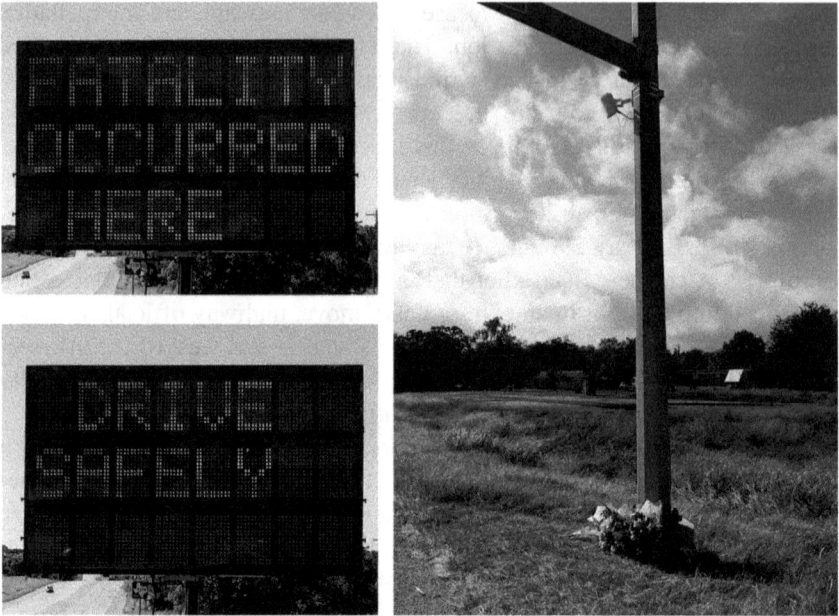

Figure 6.8.   U.S. Highway 290 @ Texas Farm-to-Market Road 2336, McDade, TX, USA, October 2016.

material objects and rhetoric to demonstrate that a road death occurred here (materializing road trauma). The shrine also shows *who* was killed there *and* demonstrates that someone who knew them, a private citizen with no invested authority to mount a speech act in the public right-of-way, has taken the extraordinary and officially illegal measure of building a shrine devoted to the person who died here, actively working-through the trauma (mourning, performing trauma). And finally, the shrine itself doesn't just sit there working for the people who built it, but addresses a larger audience as well, reminding passing motorists to drive safely (interpellating drivers and warning them) to avoid their fate.

Moreover, if we widen the frame a bit further, and notice that to the right of the official temporary sign in figure 6.9, there is another, much older, road trauma shrine, and that at the intersection where the shrine is there is also the town cemetery, with a freshly dug grave for the latest victim of road trauma in this small town, we can see that the shrine is anchoring a larger sense of collective trauma that the town of McDade is living through. And then if you are a stranger driving by, who does not know the stories the people in McDade know, but you recall, as I do, that as you drove east through this intersection on U.S. Highway 290 and FM 2336,

Figure 6.9. U.S. Highway 290-East @ Texas Farm-to-Market Road 2336, McDade, TX, USA, October 2016.

you have already seen several prominent road trauma shrines, and you will see many more if you drive even fifteen more miles on this road—or if you are reading this book, and you have noticed that several of the shrines you have seen pictured here in this book are in McDade or also on U.S. Highway 290 but in nearby Elgin, Paige, or Manor, Texas—you can see how that one single, quiet shrine at the base of that steel pole is engaged in a much larger performative enactment entirely: doing the work of showing us that each one of these crash shrines is part of an immense cultural trauma that we have only begun to *sense* and *feel* as a collective, much less *work through*.

NOTES

1 Divya Tolia-Kelly, "Locating Processes of Identification: Studying the Precipitates of Re-Memory in the British Asian Home," *Transactions of the Institute of British Geographers* 29 (2004): 316.

2 See Louis Althusser, "Ideology and Ideological State Apparatuses (Notes toward an Investigation)," in *Lenin and Philosophy and Other Essays*, trans. Ben Brewster (New York: Monthly Review Press, 2001), 85–126.

3 Michael Warner, "Publics and Counterpublics (Abbreviated Version)," *Quarterly Journal of Speech* 88, no. 4 (2002): 418.

4 See Judith Butler, *Gender Trouble* (New York: Routledge, 1989).

5 Michael Warner, *Publics and Counterpublics* (New York: Zone Books, 2002), 72.

6 Carole Blair, Greg Dickinson, and Brian L. Ott, "Introduction: Rhetoric/Memory/Place," in *Places of Public Memory: The Rhetoric of Museums and Memorials*, ed. Greg Dickinson, Carole Blair, and Brian L. Ott (Tuscaloosa: University of Alabama Press, 2010), 6.

7 Ibid., 17. Rhetoricians will recognize that this is also the central feature of what Maurice Charland calls "constitutive rhetoric." See Maurice Charland, "Constitutive Rhetoric: The Case of the *Peuple Québécois*," *Quarterly Journal of Speech* 73 (1987): 133–50.

8 Warner, *Publics and Counterpublics*, 72.

9 Gillian Rose, *Doing Family Photography: The Domestic, the Public, and the Politics of Sentiment* (London: Ashgate, 2010), 78; see also Kurt Iveson, *Publics and the City* (Oxford: Blackwell, 2007).

10 Blair, Dickinson, and Ott, "Introduction," 15.

11 See Hamilton Bean, "'A Complicated and Frustrating Dance': National Security Reform, the Limits of *Parrhesia*, and the Case of the 9/11 Families," *Rhetoric and Public Affairs* 12, no. 3 (2009): 429–60; and Theresa Ann Donofrio, "Ground Zero and Place-Making Authority: The Conservative Metaphors in 9/11 Families' 'Take Back the Memorial' Rhetoric," *Western Journal of Communication* 74, no. 2 (2010): 150–69.

12 See Ann Cvetkovich, "Public Feelings," *South Atlantic Quarterly* 106, no. 3 (2007): 459–68. See also Berlant, *The Queen of New York Goes to Washington City*; and Warner, *Publics and Counterpublics*.

13 John M. Sloop and Joshua Gunn, "Status Control: An Admonition Concerning the Publicized Privacy of Social Networking." *Communication Review* 13 (2010): 289–308. See also Daniel C. Brouwer and Robert Asen, eds., *Public Modalities: Rhetoric, Culture, Media, and the Shape of Public Life* (Tuscaloosa: University of Alabama Press, 2010).

14 See Adriana de Souza e Silva and Jordan Frith, *Mobile Interfaces in Public Spaces: Locational Privacy, Control, and Urban Sociality* (New York: Routledge, 2012).

15 Erika Doss, *Memorial Mania: Public Feeling in America* (Chicago, IL: University of Chicago Press, 2010), 50.

16 Jill Bennett, *Empathetic Vision: Affect, Trauma, and Contemporary Art* (Palo Alto, CA: Stanford University Press, 2005), 6.

17 Frances Guerin and Roger Hallas, *The Image and the Witness: Trauma, Memory and Visual Culture* (London: Wallflower Press, 2007), 11.

18 See Avishai Margalit, *The Ethics of Memory* (Cambridge: Harvard University Press, 2002).

19 See Shoshana Felman and Dori Laub, eds., *Testimony: Crises of Witnessing in Literature, Psychoanalysis, and History* (New York: Routledge, 1992).

20 See Dora Apel, *Memory Effects: The Holocaust and the Art of Secondary Witnessing* (New Brunswick: Rutgers University Press, 2002); and Andrea Liss, *Trespassing through Shadows: Memory, Photography and the Holocaust* (Minneapolis: University of Minnesota Press, 1998).

21 Ulrich Baer, *Spectral Evidence: The Photography of Trauma* (Cambridge: MIT Press, 2005), 89.

22 Ibid., 119.

23 Alberto Barrera, "Mexican-American Roadside Crosses in Starr County," in *Hecho en Tejas: Texas-Mexican Folk Arts and Crafts*, ed. Joe S. Graham (Denton: University of North Texas Press, 1997), 278–92.

24 Judith Butler, *Notes toward a Performative Theory of Assembly* (Cambridge: Harvard University Press, 2015), 59.

25 Ibid., 178.

26 Ibid., 95.

27 J. L. Austin, *How to Do Things with Words*, 2nd ed. (Cambridge: Harvard University Press, 1975).

28 See Michel de Certeau, *The Practice of Everyday Life*, trans. Steven F. Rendall (Berkeley: University of California Press, 1984), especially the chapter called "Walking in the City."

29 Butler, *Notes toward a Performative Theory of Assembly*, 176.

30 Ibid., 55, 179.

31 Guerin and Hallas, *Image and the Witness*, 10.

32 Ibid.

33 Ibid.

34 For more information about the history of the shrine and photographs of an earlier iteration of it, see Everett, *Roadside Crosses in Contemporary Memorial Culture*, 7–9.

35 For a legal analysis of the court case, see Eric B. Ashcroft, "American Atheists, Inc. v. Davenport: Endorsing a Presumption of Unconstitutionality against Potentially Religious Symbols," *Brigham Young University Law Review* 2012, no. 2 (2012): 371–89.

36 Lieutenant Lee Perry, Phone Interview with the Author, July 21, 2006.

37 Ibid.

38 See Kimberly Scott, "DATS Trucking Sets 15th Memorial Cross for UHP Trooper on Property Where Atheist Ban Doesn't Apply," *St. George News*, December 30, 2016. The official state memorial in Hurricane, Utah, is now also duplicated by a virtual memorial at both the Utah Highway Patrol website and the Troopers Association website, accessed February 9, 2017, https://highwaypatrol.utah.gov/fallen-trooper-memorial/.

39 See Peter Jan Margry and Cristina Sánchez-Carretero, eds., *Grassroots Memorials: The Politics of Memorializing Traumatic Death* (New York: Berghahn Books, 2011); and Robert Thomas Dobler, "Ghost Bikes: Memorialization and Protest on City Streets," in *Grassroots Memorials: The Politics of Memorializing Traumatic Death*, ed. Peter Jan Margry and Cristina Sánchez-Carretero (New York: Berghahn Books, 2011), 169–87.

40 "Ghost Bikes Website," accessed August 2, 2019, http://ghostbikes.org.

Chapter 7

Conclusion: Melancholy Remains

Throughout this book, I have analyzed the material, spatial, discursive, and visual dimensions of road trauma shrines, showing in detail how they work and for whom they work. As I conclude the book, I would like to more fully explore just how challenging it will be for those of us who are secondary witnesses to the cultural trauma performed through roadside shrines to begin to carry its weight. Here I want to emphasize again that the key dynamics I have identified that do this work—place-making, transference, relations of proximity, performativity, performance, and interpellation—are all *affective* forces that work through visual, material, and spatial means. They do not work at the level of representational communication, where they can be "interpreted," "read," and "understood"—but also dismissed or argued against. Instead, they work at the level of embodied, material communication, where you can add up all the elements and always feel that there is something else there unaccounted for that makes it all work—something like a traumatic rememory, which is both palpable and hard to pin down. Learning to recognize, learn from, and live with that "something else" is the subject of this chapter.

DIALECTICAL WITNESSING

To ethically inhabit the role of witness to someone else's trauma, you must attend to testimony but always be aware that the testimony does not and cannot communicate trauma directly, much less deliver the witness to a place of knowing the trauma as if it is their own. Witnessing would not happen if not for the faith that witnessing would bridge the gap between the witness and the object of witnessing. But ethical witnessing also depends upon a recognition of the persistence of that gap within the encounter. That is why I call it *dialectical witnessing*: it is witnessing that, instead of seeking to reconcile the impossible otherness opened up by trauma, lives in the space in between.

I do not claim to "know" the trauma performed and witnessed at the sites I have analyzed throughout this book in the same way that I claim to know the traumas I have experienced directly in my own life. My claim here is more contained: I have learned to see how road trauma is materialized and performed at and through road trauma shrines, and in doing my work on shrines I have learned both to carry the burden of witnessing them and to see it as a productive force. In sharing the book, I want to share that seeing *and* that burden with others as part of a larger ethical imperative to integrate the car crash into any conception and embodied practice of automobility.

When I witness the witnesses and ask you to witness the witnesses with me, I am advocating for an ethic of witnessing that seeks to form a collective sense of trauma with the goal of *bearing witness to the witnesses* so that the burden of a trauma that is *caused* by the culture's entanglement with automobiles is *shared* by the culture that benefits from it.[1] In short, because car crashes are an intrinsic part of the system of automobility that affords all of us our mobility and autonomy within the system, all of us who benefit from automobility should carry the burden of remembering that our mobility comes with a cost. That cost has been hidden from us—not through a conspiracy of silence but through the operation of automobility itself as a cultural discourse that depends on our ongoing participation in automobility as an ordinary expectation.

The ethical responsibility for any secondary witness of a road trauma shrine is twofold: first, to recognize that shrines address the public because secondary witnesses inhabit the future that the victim was traveling toward at the point of impact; and second, to witness with empathy the attempt of insider witnesses to materialize traumatic affect performed at a shrine without claiming to then know the trauma directly after witnessing it. This is a difficult challenge. To feel the loss of the individual as a loss to the collective, you must first feel that circumstances could have been different, that those smiling faces in those shrine portraits, for instance, imagined a much bigger future than they got to live, and that loss—the loss of a projected ordinary future we ourselves continue to project for ourselves every time we get behind the wheel, and even at the moment of encountering the interrupted future of the lost—has a claim on us.

But the second responsibility is even more fraught. To dialectically witness a shrine as a stranger among strangers is to work to see it not only for what it is but also for what it isn't—to feel what it is doing without imagining that after you encounter it, you know something it can't possibly convey on its own. If we focus on projecting ourselves imaginatively into the spectacular scene of the original trauma of the crash or the secondary trauma of direct loss, instead of recognizing that we are always outside it looking in, we will misrecognize projective imagination for understanding. That is, the true challenge performed by any one road trauma shrine and all road trauma shrines as a cultural technology is to stay with the shrine and learn to not "leave the site." That means learning to see shrines as scars and not memorials: they are performing trauma, not remembering it. When they ask us to witness them, they want us to stay with them, to remain in the dialectic between feeling traumatic affect and moving on.

This I something I have had to learn over time. When I began my fieldwork on shrines, I was overwhelmed by the affective intensity of the shrines I was encountering. At each stop, I would be there on the side of the road taking pictures, thinking about the shrines, thinking about the horrible scene that must have taken place, thinking about the terrible but also often beautifully indifferent scene that now sat before me, trying to sense the spaces of silence and exuberance located in most shrines. And at each stop—from the lonely ones with narrow shoulders or blind curves to the ones in urban areas buzzing with traffic—the parallel reality of the situation didn't take long to sink in: Someone died right here. Right here! And here I am taking a picture of it. There is a thin line both connecting us and separating us. What if I die here on the roadside, next to my car, taking a picture of the place where someone else died in theirs? Irony is not quite the right word to describe the feeling. But then, afterwards, I would get back in my car and drive away, left to imagine what might have happened but did not (yet).

But recognizing a potentially shared fate is only the beginning of the complex dialectical processes located in encountering trauma shrines as a stranger. For instance, in August 2003, as I was in New Mexico on my first fieldwork trip, I stopped to photograph a shrine. Part of the shrine was attached to the guardrail, and part was set just outside it. The cross carried a torn "MOM" silk flower arrangement and the name, "Martha Martinez." The guardrail had a small dent in it, presumably from the crash.

Finding myself face-to-face with a cross emblazoned with silk flowers spelling the word "MOM," I couldn't help but feel the subjectivity of it: it pulled me in and pushed me away at the same time, saying, "This looks familiar and intimate, but it really isn't—at least not to you. You have a Mom, but this is not about *your* Mom." This was constructed by someone who is radically situated and relationally located—a person who defines themselves always in relation to the person who has died. This is not only the place where a person named Martha Martinez died but also a place where *someone's Mom* died.

As a son, brother, uncle, husband, and father, I spend a lot of time thinking about this dynamic when I am photographing shrines on the side of the road. Whenever I see a Mom cross, for instance, I can't help but imagine it as a

Figure 7.1. U.S. Highway 64-East, Carson National Forest, West of Tres Piedras, NM, USA, August 2003.

shrine to my own mother or as a shrine my daughter and my son and I might put up for my wife if she were to die on the road.

This last imaginative association with my own family is not at all random, because for most of the fieldwork I completed for this project outside of my everyday interventions near home in Central Texas, my family has been right there in the car with me, a barely absent presence always at least metaphorically and sometimes literally at the edge of the frame of my photographs.

Talk about an odd space to inhabit. There we all were, driving around the Southwest, doing family road trips focused on finding places where people died. The vehicles we drove in the field became mobile places for living on the road where others had died. And my children, now teenagers, tell sardonic stories of the early years of their lives, when we all spent a lot of time on the road driving around looking for shrines for me to photograph and study while they waited with Danielle in the car.

That particular day in 2003, at the New Mexico MOM site, as I was moving around the shrine, something else caught my eye: a child's handprint in the concrete anchoring the cross. The handprint was surrounded by several names. I assume that they were the names of Martha Martinez's children, the people for whom "Mom" means *my* Mom or *my* Grandmom. I moved closer, feeling compelled to place my hand near it—for scale and for connection.

Figure 7.2. U.S. Highway 84-West, North of Santa Rosa, NM, USA, July 2006.

Figure 7.3. U.S. Highway 64-East, Carson National Forest, West of Tres Piedras, NM, USA, August 2003.

After taking this photograph, I glanced back at our car, which was parked on the shoulder a hundred yards up the road at the next safest pullout spot. My daughter, then just shy of two years old, was napping, and Danielle was sitting in the backseat next to her.

In the viewfinder, I framed a photograph with the shrine and the car together. I was instantly distracted from actually taking the picture when a car whizzed by me just a few feet from to me and I felt the car's wake flap my shirt: Would I want my family to put up a shrine if I died taking this picture? What would I hope the shrine "said" about me? To whom? What would the shrine look like? How would my shrine be related to the existing one in the same exact place? What if *they* were hit in our parked car and died while I was out of the car taking pictures? Would I create a shrine? How would I mark the space and what would I do there? What would it look like? What would I hope it said? What would I hope it did? Would I answer these questions differently if it happened closer to home, where I could visit more often? What if we all died?

I never did take that picture. Standing there on the side of the road, I knew even then at the beginning of this project that these were not simply the questions of a "researcher" studying a cultural phenomenon with a particular "methodology" from a particular "theoretical framework" or "perspective" and preparing to make an "argument." These were the questions of a person—a fragile body witnessing the cultural work of other fragile bodies appealing to me as a person as well, a person being *moved* by an affective encounter with apparently inanimate objects commemorating a stranger I would never know.

My understanding of the nature and scope of that claim of affiliation has grown since my early years with the project, of course. So has my understanding of the melancholy I have been carrying with me since 2003. As you can see from this example of encountering Martha Martinez's shrine, I first saw the claim much more literally. As a new parent at the time, I was particularly struck by encounters with not only other MOM sites but also sites that hailed me even more personally—sites memorializing dads, brothers, sons, spouses, families of three, then families of four, and of course, Roberts and Bobs. Encountering these sites induced the surreal feeling that I was experiencing some kind of alternate reality where I was seeing my own shrine from beyond the grave. I was performing a minimal form of dialectical witnessing before consciously theorizing it—recognizing a similarity that highlighted the difference between me and the memorialized person, one predicated on my knowing that they were dead: *I may or may not one day be like you, but I am definitely like you once were.*

But there is more to dialectical witnessing than that. Dialectical witnessing is not about merging the subject and object in the colonizing act of imaginative projection. It is also not about the refusal of engagement contained in the goal of "objectively" apprehending an object as a thing itself divorced from any affective encounter with it. It requires recognition of difference and similarity at the same time. Perhaps paradoxically, the closer I come to understanding what a particular shrine *does*, the farther I feel from understanding what it *means*. Encountering a shrine, such a recognition of oneness and separateness might be phrased as an imaginary address a witness like me could make to the lost person:

> Like me, you were driving on this road; unlike me, you now are no longer driving here. This shrine marks the spot where our realities diverged, but at the same time, the site opens up a space for us to converge in a witnessing encounter. By witnessing your absence through witnessing the performance of traumatic memory and mourning at the site, I see for the first time that you were once present. Now, however, I see your presence only in the form of a shrine that marks your absence. Your shrine has made me aware of what we share, but also what we don't. By witnessing the shrine, I not only see and feel that you once were here on the same road but also that you are no longer here on the road with me. While you are absent, I am present, but I would not be present here now at this shrine if your absence were not made present by the shrine—if your shrine had not called me into being as part of your shrine's public.

Such an encounter is not possible without the shrine's ability to interpellate drivers as part of the "knowing motoring public." And a shrine's ability to interpellate is dependent on its ability to mark and mediate presence and absence and similarity and difference in a particular time and space. But even then, witnessing doesn't just happen. Shrines address publics and demand witnessing, but cannot ensure it. Thus, it has been one of my aims in this book to show not only how and why witnessing is demanded and necessary but also why it always also partial and problematic.

ENGAGING THE "SOMETHING ELSE"

Unlike the rest of the book, where I have moved around considerably in terms of space and time to show a wide range of different shrine sites, for this last intervention I would like to focus extensively on a single site near Austin that I have been engaging for nearly ten years.

On February 28, 2010, four young siblings from Houston, Texas—Paul Gonzalez, aged nine; Noel Gonzalez, aged eight; Angelina Gonzalez, aged six; and Aaliyah Ann Gonzalez, aged five—died in a car accident just outside the small town of Paige, Texas, about forty miles east of Austin. Even hearing these simple facts, you can imagine how terrible the scene must have been and how intense it all must have been for the family that remained to absorb the trauma of such a massive and tragic sudden loss. But this book has not been about how strangers like us feel something when they imagine absent stories about absent people. It has been about understanding how affect works when those absent people are made present not through direct representations but through performative enactments in the spaces and objects that make up road trauma shrines.

A road trauma shrine devoted to the four Gonzalez children has existed at the southeast corner of U.S. Highway 290 and Texas Highway 21 since early 2010. Because the location is not far from my home, I see the shrine often in the midst of my ordinary travels and I have stopped to photograph it each time I have noticed a significant change to the site. Because the shrine has changed a lot since I first saw it in April 2010, and because my relationship with the site has continued to deepen with each visit to it, I would like to focus on it here as a final example that shows how the visual, material, spatial, and temporal dimensions of road shrines do the cultural work they do, but also the work they do not do.

I first photographed the Gonzalez site in April 2010, less than two months after the crash. Then, the newly built site featured four wooden crosses set in concrete (see figure 7.4), with the crosses in the following order from left to right from the perspective of the road: Paul, Angelina, Noel, and Aaliyah.

Figure 7.4. U.S. Highway 290-East at Texas State Highway 21-North, Paige, Texas, USA, April 2010.

The two crosses for the boys were painted blue, and the two crosses for the girls were painted pink. There was a different colored stuffed rabbit placed at the top of each of the four crosses: a gray one holding a basketball for Paul's cross, a purple one for Angelina's, a brown one holding a soccer ball for Noel's, and a pink one for Aaliyah's. Each boy's cross also included a toy race car, while each girl's cross featured an identical small stuffed baby doll attached so that it looked as if the rabbits were holding the dolls.

Near the middle of the site, set back a bit behind the crosses, was a shepherd's hook plant hanger wrapped in silk ivy and flowers and holding a large white stuffed bear. The bear had a bow tied around its neck and a card with a product name and care instructions attached to one ear like an earring. Writing on the card read, "From: Tia Angie & fam." Writing in the same handwriting on the bow listed the children by their more intimate nicknames and read, "We Love You & Miss You." This message from the children's aunt is the only identifiable indication of who might have built the shrine, placed any specific items there, or otherwise interacted with the site. It is also significant that the stuffed toys at the site in 2010 were all baby-proof: the white bear is from the Baby Ganz line of toys designated safe for children from birth on, and the baby dolls attached to the shrine have stroller clips on them. This was something I only recognized as the central toy at each cross later was replaced by something meant for older and older children each time.

Here, less than two months after the crash, the site already showed a remarkable clarity of design. Even here, though, there was a direct reminder of the original trauma present at the shrine: the skids marks on the shoulder, the damage to the guardrail, and the red flag markers left there from the original crash investigation (see figure 7.4). Judging from the two parallel skid marks bisecting the guardrail, the many marks on the guardrail itself and its supporting wooden poles, and a broken reflector pole on the ground in between the guardrail and the shrine, it appears that the vehicle carrying the children jumped the guardrail and knocked out the pole before landing where the shrine has been built. At the shrine site itself, about ten feet past the guardrail, the grass was trampled and the dirt disturbed by the crash or by the process of removing the crashed vehicle from the site. Several red marker flags were still in place from the police investigation that followed the crash. Some of the red flags marked the final resting place of the four wheels of the vehicle involved in the crash. The shrine was built right in between two of the red flags, at either the front or back of the vehicle's footprint. The most important observation here is that like many shrines, the shrine site was overwritten on the crash site and that the crash site itself maintained an alternate but no less material inscription of the police investigation early on in the life of the shrine, with Aaliyah's cross on one end serving as the pivot point connecting the two sites, one inscribed upon the other.

By the second time I photographed the site, in February 2011, just before the first anniversary of the crash, the site had undergone a major revision (see figure 7.5). The flower crosses had been moved to the back of the wooden crosses, revealing that each cross carried a bronze placard carrying the name, birthdate, and death date of each child. The whole site now featured a black plastic border that was filled with white gravel and ringed with solar lights to light the site at night. The red marker flags were nowhere to be seen by then.

More important, the stuffed rabbits were all gone, replaced for the boys with full-scale American footballs and replaced for the girls with new, much larger life-size baby dolls that leaned against each girl's cross as if sitting in a chair. Through the revision, the gender coding of the four crosses had

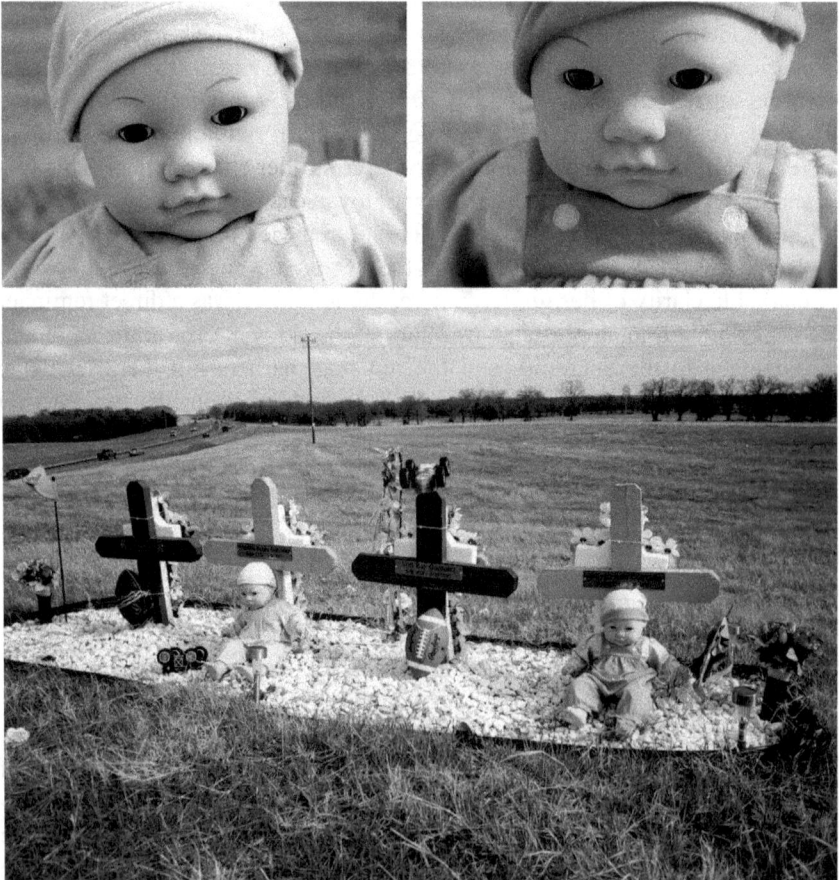

Figure 7.5. U.S. Highway 290-East at Texas State Highway 21-North, Paige, Texas, USA, February 2011.

become even more starkly elaborated. This was an intriguing change from smaller-than-human scale to human-scale objects for all crosses, and a shift from anthropomorphic animals to nonhuman objects for the boys and a shift from anthropomorphic animals to actual human forms for the girls.

The site now seemed to be dominated by the presence of the two new dolls. The salience of the dolls made the footballs feel that much stranger: if the central objects placed at the base of each cross were functioning as symbolic proxies for the lost children, it made sense to have the dolls sitting upright like little people, but the same could not be said about the footballs, which were placed in the same position—particularly Noel's football, which appeared to be leaning back in its chair like a person reclining (see figure 7.5).

Faced with this stark demonstration of how traumatic loss can be given form and anchored to the ground, I felt connected to the site more than ever before. At the time, I had just begun to develop my central argument about the way that road trauma shrines territorialize memory by transferring the life lost in an automobile crash to the life lived by the shrine itself placed on the roadside. I was noticing that this process of transference replaces the interrupted life with a new life that is allowed to take its course as the lost person's life was expected to before it was cut short by the crash. One of the things I was beginning to analyze was the way that shrines acting as proxies actually *age* on the roadside, extending the life of the lost person into a future that had been interrupted and negated by the crash. After watching this particular shrine change dramatically over the course of a year, I projected a near future time when I would return to photograph the shrine to show how the dolls had aged in the weather, as I had documented at other sites throughout the American Southwest by then for several years.

Just two months later, in April 2011, I noticed that the site had been revised again—less dramatically than the last revision but nonetheless in significant ways. The site had been cleaned up a bit, and Paul's racecar, which had been on the ground in February 2011, had been replaced on the top of his cross. More important, all four crosses featured a new, equally gendered, object. Even more significant than the fact of the additions was the manner in which they were added to the crosses: they were attached to the existing central object "sitting" at the base of each cross, reinforcing the function of the original object as a proxy for the lost child their cross represented. Again, this was much more literal with the girls/dolls than with the boys/footballs. Each baby doll now appeared to be holding a smaller baby doll of their own, pink for Angelina and lavender for Aaliyah—just as the rabbits had done in the April 2010 iteration (see figure 7.6).

Placed as they were, they seemed to be offerings for each child's proxy— where the sitting dolls represented the lost girls, and the new dolls were singularized gifts for the proxy girls. This sense of the new encased dolls

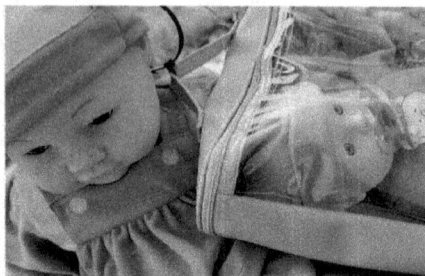

Figure 7.6. U.S. Highway 290-East at Texas State Highway 21-North, Paige, Texas, USA, April 2011.

functioning as gifts was reinforced by the fact that each of the sitting dolls was also now holding a matching plastic butterfly in their other hand. As with the original placement of the footballs at the base of each boy's cross, here again the representation schema was much more symbolic: Each football now appeared to be "holding" an identical Nerf gun, as if the footballs represented the lost boys, and the guns were gifts for the proxy boys (see figure 7.6).

The year went on, and I watched the shrine as I drove by it many more times. There it sat, doing what I was now learning to expect road trauma shrines to do: standing there in the sun and rain, living its life, faithfully standing in for the lost children, keeping them alive socially even as the children were no longer embodied themselves, trying to get us to see that their loss had something to do with us. The site turned into a reference point, a landmark on the way through my everyday life as both a driver and a researcher.

And then, in December 2012, I noticed another dramatic development. Actually, it barely registered as I drove by at my regular full speed, but once my mind caught up with my eyes, I instinctually jammed on the brakes and swung my car around to park in what was now a familiar pullout spot just south of the intersection. Approaching the site, I confirmed what I had only partially apprehended moments before as I drove by the site: there were now different dolls at the site (see figure 7.7). Gone were both the sitting dolls and the smaller dolls they had seemed to be holding, and new dolls had taken their place. The boys' footballs also had been replaced with soccer balls, and the Nerf guns that had been attached to the footballs now lay on the ground beside each boy's cross. A basket of silk flowers was now hung on each cross—blue and white for the boys and pink and white for the girls. The white gravel border was still there, but obviously was overgrown.

When I approached the site and saw the new dolls up close, I froze. The dolls were clearly no longer supposed to be babies. They had the jaunty, coordinated outfits and stylish haircuts of much older girls. There it was: if the shrine was a stand-in for the lost children, whose lives were interrupted, the shrine itself was not only enduring a longer life in place of the lost children, showing signs of aging as it was exposed to the elements, but also more literally *growing up* as it did!

Where the baby dolls looked at me with big, expressionless baby eyes, hailing me as a grown-up, drawing me in to them as a potential caretaker, these new "big girl" dolls looked out with confident smiles on their faces, ready to take on the world, not seeming to ask for anything other than recognition. And yet, the girls they stood in for, Angelina and Aaliyah, the girls who would have now been almost three years older, who might have one day looked forward to a future filled with matching outfits and confident though fragile expressions, were gone. Their lives, once filled with ordinary rhythms

Figure 7.7. Aaliyah Doll (Left) and Angelina Doll (Right), U.S. Highway 290-East at Texas State Highway 21-North, Paige, Texas, USA, December 2012.

and punctuated by everyday affects, were gone, erased in the blink of an eye that day in February 2010. But they remained here in this new materialization.

Staged this way, both the family's loss and the family's work to negotiate it became not only visible to me but also palpable. A moment later, still shocked by the poignancy of what I was seeing, an intense wave of sadness engulfed me, and the sadness stayed with me the whole time I photographed the site. Melancholy certainly followed me home that day, as I opened my MacBook to compare my new photographs to the previous photographs I had taken of the site. It only got deeper from there as I connected the dots from the first iteration of the shrine to the latest one.

That is when I first became intrigued by the changes to the boys' proxies. Where the shift from baby doll to big girl doll was a literal materialization of the way the shrine continued to grow beyond the lives of the people it memorialized, the replacement of one type of sports ball for another materialized a change, but one only abstractly read as "growing up." If I had seen the balls replaced at another shrine without the dolls next to them, or hadn't recognized that the earlier footballs appeared to be sitting up (and later holding things), I might not have seen the change from one ball to another in anthropomorphic terms, but staged as it was here, the change in sports balls seemed consistent with the ways that children's interests and activities change as they grow up, and how the grown-ups in their lives play a role in fostering (and resisting) that.

For the next two years, I drove by the site many times, and stopped a few times to photograph the site as the shrine was subjected to the rhythm of the seasons and the slow but inexorable process of decay. The next time I photographed the site, in May 2013, the surrounding grass and wildflowers almost had swallowed the site. The "big girl" dolls had begun to mildew, making it appear that they were getting sun freckles on their faces. Their hair had begun to unravel a bit. The doll at Aaliyah's cross had already begun to be bleached in the sun.

Figure 7.8. U.S. Highway 290-East at Texas State Highway 21-North, Paige, Texas, USA, May 2013.

Figure 7.9. Angelina doll (Left) and Aaliyah Doll (Right), U.S. Highway 290-East at Texas State Highway 21-North, Paige, Texas, USA, May 2013, January 2014, and August 2014.

By January 2014, the summer grasses around the site had been mowed by the state maintenance crew, but some of the grasses remained to cover the white gravel bed border and even to grow through the dolls' clothes. The mildew "freckles" now covered almost all of the dolls' exposed skin. The hair on the doll at Aaliyah's cross was almost entirely sun-bleached, giving the uncanny feeling that she was going gray there on the roadside, much like that Tasmanian Devil doll I had seen years earlier in New Mexico. The boys' soccer balls remained in place and inflated as they, too, began slowly to fade.

The doll at Angelina's cross was showing its age in a more violent way: some time that year, an animal had chewed through both of her hands.

In August, 2014, I noticed another dramatic development: the slowly aging "big girl" dolls remained, but they were now joined by two new dolls strapped to the top of each Gonzalez girl's cross. The big girl dolls continued to age, with their skin more covered with mildew spots than not, and discoloration creeping across their faces. Aaliyah's doll's hair had entirely turned white. The two boys' crosses also included a new addition as well: a blue water gun for Paul and a green water gun for Noel. By this point, the older Nerf guns also had been swallowed by the grass growing at the base of each boy's cross.

Figure 7.10. Angelina dolls (L) and Aaliyah Dolls (R), U.S. Highway 290-East at Texas State Highway 21-North, Paige, Texas, USA, August 2014.

Figure 7.11. Angelina Dorothy detail, U.S. Highway 290-East at Texas State Highway 21-North, Paige, Texas, USA, August 2014.

Figure 7.12. Full Shrine (Top), Angelina Dorothy Doll (Left), and Angelina Doll (Right), U.S. Highway 290-East at Texas State Highway 21-North, Paige, Texas, USA, September 2016.

Unlike earlier dolls at the site, which were always different from one another and tied directly to base of the crosses, the two new dolls were identical to one another and tied at the top of the cross. Like the smaller gift dolls placed there in 2011, these new dolls were encased in their original plastic packaging. Also distinctive was the fact that the dolls now were not representations of generic fictional "babies" or "big girls" but of a well-known character from American popular culture: Dorothy, from *The Wizard of Oz*. More specifically, the Dorothys were merchandise from the 2013 animated movie *Legends of Oz: Dorothy Returns*.

The last time I photographed the site, in September 2016, most of the items placed there by August 2014 remained, but had clearly aged. The flower baskets, there since 2012, were now gone. The dog and cat ornament, there since 2011, was now rusted. The white gravel and border, placed there in 2011 and barely visible by 2012, was now completely overgrown. Noel's green water gun from 2014 remained on the ground, and the entire site was shrouded by tall grass, making the shrine difficult to see from the road.

The Dorothy dolls had fared better in their plastic cases than the previous dolls had in the open air, but even these dolls had faded a bit and the packaging itself had faded considerably. Now six years old, the crosses themselves were showing signs of aging as well.

If my photographs and words have conveyed anything of these scenes to you, maybe you are now feeling something as you read this. Maybe, like me, as you see the shrine, you imagine the person who maintains this shrine behind the scenes, probably a parent, or an aunt or uncle, who has been performing this grief work. You can imagine how they plan each revision—visualizing the design of the site, going to the store, finding just the right dolls to singularize and place at each girl's cross, gathering the tools and materials necessary to install them, driving to the shrine site, connecting the new items to the crosses strongly enough to withstand the elements, gathering the replaced dolls, driving home afterward. With empathy, you can imagine what they must be feeling as they do.

But that is clearly something in our minds, something imagined and projected onto the site. There is plenty at the site itself and in these photographs that seems to trigger sadness. There is *something there* in the objects and how they are assembled, not only in the stories behind the scenes. There also is something *else* there—something there at each moment pictured here, and something also there in between the moments as you look at pictures showing how the site has evolved in the years I have been photographing it.

There's something poignant about any road trauma shrine, where the tragedy commemorated becomes apparent in a different way than it does when you read a news story or an obituary, see a gravesite, or visit a mausoleum. There's some palpable force of undeniability to the form of a shrine itself: right here, on this spot, someone was killed in an automotive accident, and right here, someone else is working-through their traumatic grief by building and maintaining a shrine at the same spot. And when the shrine literally stages an ongoing life for the ones lost at the exact location where that trauma occurred, transferring their lost life into the life of the shrine itself and the objects within it, a shrine can become the location of the experience of vicarious trauma and sometimes intense affect for the strangers who witness them. Seeing this shrine not only appear but also continue to evolve was poignant to me because it *performed* a trauma that might otherwise go unnoticed—either because it was not known to me or because it hadn't been given a material form so that I, a complete stranger, could know about it.

But it was more than that.

There was something else—something I felt and knew each time but couldn't quite put my finger on, much less put into words. That "something else" is the melancholy about remains that remains in road trauma shrines, where affect is placed and temporalized but never quite contained, spatially, temporally, or culturally.

It seemed so natural each time it happened to me—my intense feelings of melancholy, that recognition of poignancy—something akin to what I had experienced many times before at other shrines, including the site where we opened the book. But here, with this apprehension of a transferred proxy unfolding in life stages, it was even more intense. But what is the nature of this particularly melancholic kind of affect? What does this kind of affect *do*, and what do we *do with* it once it registers with us?

Melancholy is always structured as an excess of affect. Melancholy remains. It is a feeling that *persists*, but the things that trigger it—that make it persist—also are *leftovers*. They are a bridge between realities. They become a remainder. I am melancholy about remains, and that melancholy remains. It is persistent. It stays with me, and it continues to be something else left over after an encounter with someone else's remains.

REMEMBERING STRANGERS, DRIVING INTO THE FUTURE

As I have been arguing throughout this book, strangers encounter road trauma shrines in everyday landscapes without knowing the people memorialized, without being "inside" the micropublic of insider witnesses that maintains a social presence for the victim by performatively commemorating a specific life lost, but being contained inside a different macropublic: outsider witnesses in the form of the knowing motoring public made aware of road trauma as a cultural trauma, drawn together by an implicit assertion of affiliation—an assertion that *their* trauma is *our* trauma. But what exactly are we to do when we recognize that we are being addressed as if a stranger's trauma is *our* trauma? It is one thing to recognize a rememory when you see or feel it. It is quite another to know what to do with it once you do.

To help explain this dynamic, I want to think through road trauma shrines as a form of what Rico Franses calls "stranger-memorials."[2] For Franses, a stranger-memorial is a distinctive collective memorial form that commemorates absent people for a public that does not have direct, personal knowledge of the people being memorialized. The stranger-memorial attempts to bridge this gap by staging a relationship between the stranger and the memorialized dead in the present, but it never quite accomplishes the memory work it is designed to do. The reason it doesn't is directly relevant to the questions I am exploring here in relation to road trauma shrines, so I want to look more closely at Franses' argument before I bring us back to the question of how roadside shrines place—and don't place—memory and affect in shared landscapes to materialize a cultural trauma.

Franses developed his idea of the stranger-memorial in his 2001 article titled "Monuments and Melancholia," where he concentrates on two familiar case studies of contemporary collective memorial forms in the United States: the Vietnam Veterans Memorial and the AIDS Memorial Quilt. Both of these memorials emerged contemporaneously to the losses they mark; that is, unlike older memorials to long-lost people, including Confederate memorials, where the passage of time ensures that all visitors are necessarily strangers to the loss the site commemorates, the Wall and the Quilt were created to mediate the trauma of loss experienced by people who had a direct stake in ensuring the public commemoration of otherwise private losses contemporaneously, but in a public place where strangers also would encounter them. Not long after Franses wrote this, we watched this process unfold again with the development of the National September 11 Memorial and Museum. Like road trauma shrines, these memory forms have two main constituencies—insider witnesses who know the dead intimately and mourn them with and through these memorials and a much larger public of outsider witnesses who is interpellated as if they should witness that known loss and make it their own as part of some larger collective.

As Franses says, though, there is a problem: while the first group already knows who and what is lost, strangers, who have only the memorial, are left outside looking in—left witnessing the material signs of traumatic absence but not really knowing what is not there. This fact is particularly stark at the Vietnam Veterans Memorial, which has the evocative physical form of a scar rising from the landscape, but uses only names to evoke the lost people it

commemorates. Strangers are capable of reckoning with the aggregate loss—
especially in this particular memorial, which implicitly recuperates the individual losses into a larger national loss—but not the specific, unique losses.

The AIDS Quilt is a collection of individual quilt panels that unfold
according to their own design; most contain a large number of different
applied objects and representations in addition to the name of the person
memorialized. Here, Franses argues, the "underpinning logic seems to be that
the more objects included in each panel, the greater the reference to the dead,
and the stronger the evocation will be."[3] At first glance, then, it might appear
that by showing and evoking more of the person lost, the Quilt resolves the
problem of the stranger-memorial, which is that one cannot remember—or
forget, for that matter—what one has never known. But Franses argues that
this is merely the "lure" of this more elaborate kind of stranger-memorial.[4]
Indeed, Franses writes, "The difficulty it faces is how to induce affect upon
strangers for a loss that . . . never happened to them; for the effective memorial is one which must first convince strangers that a loss has indeed occurred,
and that it is their loss." However, "rather than overcoming the sensation of
non-acquaintanceship, . . . the quilt accentuates it."[5]

To elaborate on this important insight, Frances turns more specifically to
Freud's influential essay, "Mourning and Melancholia."[6] There, Freud theorized that the function of mourning was to free the bereaved from attachments
to the lost object of their grief. Failing to do so would lead to melancholia—a
pathological attachment to the object of grief and loss with its consequent
diminishment of the bereaved person's ego. Freud theorized that melancholia
establishes an "*identification* of the ego with the abandoned object," where
"the shadow of the object [falls] upon the ego." In mourning, the bereaved
may eventually become free to attach to a new object, but in melancholia, the
abandoned object gains a "special agency" over the ego, and "object-loss [is]
transformed into an ego-loss."[7] As Freud argues, this "substitution of identification for object-love" has two important consequences: not only does it
diminish the self-value of the bereaved, but it also overvalues the object that
is lost but cannot return.[8] The result is a desire for reunification with the lost
object that is as strong as it is impossible to achieve. Freud writes that melancholia "behaves like an open wound, drawing to itself cathectic energies . . .
from all directions, and emptying the ego until it is totally impoverished."[9]

The goal of Freudian-influenced grief therapy was to create a space where
the patient could work through their grief in a stimulating but safe transferential therapeutic space to "break the bonds" of grief attachment to lost, absent
objects to allow the patient to resume a "normal" life, which would entail
being able to attach to new present objects. This model of grief therapy
was dominant for decades in psychoanalytic theory and practice, and has
significantly affected the dominant cultural concept of grief and mourning as

a temporary, pathological state that should be overcome either through therapeutic interventions or through willful forgetting. Indeed, especially after the model was fleshed out even further and popularized in 1969 by Elizabeth Kübler-Ross in her theory of the stages of grief—denial, anger, bargaining, depression, and acceptance—the dominant model of grief in popular culture has been focused on working-through grief in a certain standard process and working toward the goal of "closure" and the avoidance of "complicated grief."[10]

In his later work on mourning, informed by his own personal experiences with grief and loss, Freud himself theorized a more continuous form of grieving and mourning and a constant slippage between mourning and melancholia. Likewise, as we have already seen in our analysis of grief performances at road trauma shrines, contemporary theories of death and bereavement have focused less on "getting over" grief and more of a "living with" the "continuing bonds" of grief.[11] In contemporary bereavement theory, we see a definition of working-through as a punctuated but continual lifelong process. This is the model of grief that seems to be materialized in stranger-memorials, especially trauma shrines and road trauma shrines: while the rhetoric of closure is still present in the verbal and written messaging, the cyclical renewal of many public trauma shrines also indicates a more continuous conception of the presence of the dead in the lives of the living and a refusal to see their deaths as contained or finished. Road trauma shrines thus are a material example of how mourning is *made to keep being present* in contemporary public spaces. There, these acts of ongoing mourning are witnessed by strangers, who are always already positioned differently to both the material form of such memorials and the processes surrounding them.

Franses argues that the main affect produced in encounters with stranger-memorials is melancholia, where "one is possessed by the structure of a loss, but the place that ought to be occupied by the lost object is vacant."[12] The problem is that in the stranger-memorial, "everything happens in reverse. One is introduced to someone only after he or she dies. One grieves for someone one never knew. One mourns an object one never possessed."[13] Where acquaintances of the lost dead have the potential to heal through mourning, "such mourning is impossible for strangers." Put simply: "One cannot mourn (in the technical, psychoanalytic sense) persons to whom one has been introduced only after they have died. . . . He or she is always-already a stranger."[14] Franses argues that witnessing the prolific communication of intimate details in the AIDS Quilt actually *extends* the power of melancholia: in Freudian terms, apprehending the affect of each memory object in the Quilt panels "requires a larger slice of the ego to match it, producing greater melancholia."[15] As Franses puts it, "The stranger-memorial, we might say, operates in the mode of deceit. It tricks one into believing that one has suffered a loss, and then provokes melancholia as the reaction to the event."[16]

Which brings us back to road trauma shrines. As Michael Warner argues, "A public is a relation between strangers."[17] Like other stranger-memorials, road trauma shrines interpellate strangers, addressing them as if they are part of a public. Because shrines do so in the everyday landscapes of automobility, however, they further define that collective as the "knowing motoring public," a collective that, especially in the United States, is resolutely dedicated to living with and through cars. What is "placed" in a shrine then is not precisely a specific communicable *memory of someone*, but the *material evidence that someone else is experiencing trauma in front of us*. That lost object, that *someone who someone else remembers*, is forever suspended elsewhere, beyond the stranger's ability to know. What remains, as Kathleen Stewart says of other worldings, is the possibility of feeling that "we're in it together, whatever it is."[18] It's a small opening for collective affect, but an opening nonetheless.

When you drive by a shrine and notice it, you are brought into its web of transference, where you can experience collective affect. Of course, saying that affect is collective does not mean that it is uniform. This is where I think Franses overstates his argument. Where Franses emphasizes melancholia as the main response to stranger-memorials, I would argue that melancholia is instead actually the *best-case* scenario response—the closest we might come to feeling anything that such memorials presume us to feel. I suspect most won't even feel that.

As I have argued throughout this book, road trauma shrines are places where affect collects and is experienced, but shrines elicit a wide variety of affective responses.[19] Many people see them as effective memorials, tributes, or warnings, but others find them kitschy and sentimental, and many find them morbid and "creepy." Others may see their utility, but are confused or offended by them because they consider them "out of place" in the public right-of-way. What all of these responses share is a kind of emotional intensity that exceeds the words used to express the perspective: people seem to have strong feelings about roadside shrines, whatever those feelings are. This intensity is another sign of that elusive "something else" so characteristic of affective relationships to landscape. To say that shrines place affect is to say that they are the location of affective responses, not any one *particular* affective response.

Even the public response most aligned with the way shrine builders present them—as poignant memorials to lost loved ones—is not as simple as it may first appear. If, like Franses, we analyze the responses to stranger-memorials that focus on sadness, poignancy, or melancholy, we see that they produce a melancholy not based on a condition of *knowing* but on *not-knowing* the loss the shrines represent. As Franses would argue, the result would be not only melancholia, but *shared* melancholia. Importantly, what the public of a stranger-memorial shares is not collective *memory*, but collective *affect*—a collective felt sense that "something else" is there, even if you "can't put your finger on it."

We may recognize that a loss has occurred, a loss that is as impossible to know as it is real to perceive, but it is a loss that can never actually be mourned directly by strangers. This leaves the stranger at most holding onto their own melancholia toward a generalized sense of loss, with no resolution possible except a recognition that their own condition of grieving a loss not their own is at least *shared* by the other strangers who encounter the stranger-memorial next to them as well.

Crucially, *this* is nature of the collective affect that can be shared by even the most empathetic witnesses of the stranger-memorial: the shared experience of witnessing its frustrated attempt to communicate the traumatic memories it performs, and the shared condition of being a stranger alongside other strangers in relation to someone else's loss when someone else is asking you to see the loss as your own. In the end, in encountering a road trauma shrine, the gap between those who *know* the loss and those who *know of* the loss remains unbridgeable. The stranger can feel something, and the stranger can feel implicated, but they can never in any meaningful sense of the word *remember* what they have never known. They can, however, recognize that they are not alone in this feeling of being located in a dialectic.

I have been witnessing shrines for long enough to know that while shrines may look like windows onto some other reality, I have come to see them as heterotopic, heterochronic mirrors held up to the places and people around them. Both as a scholar and as an empathic person, I may desire a bridge to understanding the losses commemorated and performed by road trauma shrines, but where I may seek a bridge I always find a gap.

Crucially, however, that gap is not an empty void. As I hope I have demonstrated throughout the book, it is a space that performatively enacts an intersubjective space of what Kathleen Stewart calls "we-feeling." It is a different register of affect and trauma than that felt by insider witnesses. It is the affect that strangers can feel when encountering another's intimate affects, and doing so alongside other strangers, which creates its own kind of affect: the feeling of being "in" something with others, even if we are not feeling the same thing.

As John Wylie argues, affective landscapes can be full and empty at the same time; they can entail "a simultaneous *opening-into* and *distancing-from*."[20] My work has been all about *staying there in the gap long enough to feel something and understand what I am feeling*. When I do, I realize that this dialectic of opening and closing is exactly where the affect in encounters with stranger-memorials like road trauma shrines is located: in the gap between what we feel and what we know we don't know, which creates an intersubjective space for recognition. It creates a space for recognizing a shared sense that other people's ordinary trauma has something to do with us. While that shared sense is vague and easy to disavow—and certainly not the same as the deep feeling circulating through the shrines for intimates—it is *something*.

FEELING MELANCHOLY TOGETHER

Karen Wells has shown that trauma shrine sites experience three main phases of activity: shock, melancholia, and mourning.[21] In the first phase, shock about the tragedy dominates, with emotions raw and unfocused. In the next phase, melancholia, the tragedy is broadened out to a larger public, making a public claim that the death should not have occurred, and asking for justice. As Wells observes, in this phase, "the interpretation of events and their significance for wider political questions of the unequal spatial and social distribution of violence are at the forefront of commemorative practices."[22] Finally, in the mourning phase, the shrine either is replaced by a more permanent memorial structure or disappears, which "closes down the political space opened up first by tragedy and then by melancholia."[23] Wells emphasizes that the phase of "melancholic memorialization" has the most potential for collective political action, because as the shrine opens itself to the world, "the meaning of the tragedy is located on the border between private grief and public justice."[24] Wells argues that "because of its refusal to foreclose the political signification of loss," melancholia can be a powerful force within collectives.[25] Because melancholia is "a refusal to let go of the lost object," melancholic memorialization keeps lost objects visible and material in the social landscape so they can serve a purpose.[26]

However, the outcome of such a process is anything but determined. Some trauma shrines, such as the early shrines for the Oklahoma City bombing, Ground Zero, or even Princess Diana, become the focus of collective action that then is translated into a collective movement to build a permanent memorial that can allow the public to move from melancholia to mourning, with all the costs and benefits that entails. However, this is a rare outcome seen only in the most prominent public traumas, and is especially rare for a road trauma shrine, reserved for bus accidents and multicar collisions that kill a large number of people at one time. Some trauma shrines are removed by institutional authorities tasked with policing public space. Most trauma shrines simply age into oblivion. Their disappearance then could be read as a sign either of neglect or of "successful" mourning, but either outcome is a lot different from morphing into an institutionally established permanent memorial.

Indeed, given the fact that road trauma shrines hardly ever develop into permanent memorials, we might say that almost all private road trauma shrines only go through two of the phases Wells describes: shock and melancholia. Restated in terms of the language I am using here, road trauma shrines take the form of either wounds or scars: they focus on either the trauma of a crash or the trauma of working-through a crash. They never achieve the institutionalized status of a historical marker that would not be a site of active working-through but more literally a memorial: a *reminder* that working-through once occurred at the site but it is no longer visible as a scar, much less a wound. Such a memorial would then not technically be a road trauma *shrine* any more, but a roadside *memorial*, whose function was to remind people about something that is absolutely absent—both the initial road trauma *and* the road trauma shrine as evidence of the active working-through of the event at the site.

For their lifespans as shrines at least, road scars have the *potential* to interpellate a public that would see the tragedy as its own. The fact that they rarely succeed in this is indeed the source of the melancholy of road trauma shrines. They remain there on the roadside, with the potential to gather a public, where mostly they are what Wells calls a material "statement of presence, a refusal of erasure."[27] But their failure to gather an external public seeking justice—their precarious location between what Wells calls "the simultaneous promise and failure of the dead to speak"[28]—is to me exactly what makes them poignant, not their successful conjuring of loss.

When you drive by a shrine and notice it, it can "trigger" your own memories of road trauma. It can "make" you slow down. It can "cause" you to feel empathy, to perceive that the site is "poignant." Literally, the shrine does nothing but, nonetheless, *things happen*. What happens is that you are brought into a melancholic web of transference, where you feel affect, but have no known object to attach it to. As with individual melancholia, a shared melancholia is persistent exactly because it is not attached to a known object of grief, leaving it unsettled, undetermined, unprocessed. Shrines perform traumatic memory in the present, reminding us that losses occurred, but they can't, by themselves, move us past that *sensing* of loss to a *knowing* of loss. Instead, they open up a dialectical space, a space for recognizing loss and recognizing the impossibility of ever fully "healing" from that loss. That space is not only invitational but also ethical: it demands engagement, even if it will be frustrating and frustrated.

With mourning endlessly deferred for most witnesses at stranger-memorials, therefore, melancholia remains. Melancholia not only endures but also adheres to the objects it materializes. It *stays there*, functioning simultaneously as an *eruption* of the past into the present and an insistent *interruption* of the present. Melancholy remains to be seen, meaning that it persists so that it can be witnessed, but also meaning that it also might not exactly be visible, might not be possible to be seen.

But if melancholia is what strangers share, what do we do with it, and what does it do to, for, and with us? Is it simply a vague but infinite sadness, or is it potentially a productive force, even a social agent? This is a question taken up directly in an edited volume called *Loss: The Politics of Mourning* (2002), which explores "the numerous material practices by which loss is melancholically materialized" as a productive social, cultural, political, and aesthetic force.[29] In this book, presciently written just before but released just after the events of September 11, David L. Eng and David Kazanjian note that "to impute to loss a creative instead of a negative quality may initially seem counterintuitive," but they argue that "melancholia offers a capaciousness of meaning in relation to losses encompassing the individual and the collective, the spiritual and the material, the psychic and the social, the aesthetic and the political."[30] Instead of dismissing melancholia as a pathology to be avoided or moved past, then, we might do better to instead steer right into it so that "a better understanding of melancholic attachments to loss might depathologize those attachments, making visible not only their social bases but also their creative, unpredictable, political aspects."[31]

In the same volume, Judith Butler argues that loss is constantly "bringing bodies to the foreground" in the "anachronistic" form of "voiceless mimes" that both register loss and "become the means by which that loss is registered."[32] Loss performs a past trauma whereby the "past is not actually past in the sense of 'over,' since it continues as an animating absence in the present, one that makes itself known precisely in and through the survival of anachronism itself."[33] Given that the loss itself remains and the remains of the loss remain, "we are, as it were, marked for life, and that mark is insuperable, irrecoverable."[34]

However, this is not simply a burden to simply be endured—a life sentence, a terminal diagnosis. Faced with irrecoverable trauma and loss, melancholy remains not only our condition of being but also our condition for *being together*. Butler argues that a public that recognizes itself through loss is also always bound by loss; likewise, a place of loss is "a place where belonging now takes place in and through a common sense of loss." At such a place, "loss becomes condition and necessity for a certain sense of community, where community does not overcome the loss, where community *cannot* overcome the loss without losing the very sense of itself as community."[35]

The problem is that the "community" that I am calling the *knowing motoring public* does not yet think of itself as a community, much less a community bound by loss. The people building shrines on the roadside are doing their part, but now it is up to the rest of us to take up the implicit and sometimes explicit challenge they make to the culture. We cannot do that unless we understand what shrines want from us. What they want is for us to see their loss as our loss, as the material evidence of a cultural trauma. That is not an easy thing for a stranger-memorial to do. I do not believe road trauma shrines have been as effective as stranger-memorials as they have been at serving as what we might call, after Franses, "intimate-memorials." That is, road trauma shrines are addressed to strangers, but I do not think they are (yet) effective at making the individual traumas they mark and perform feel like part of a cultural trauma. I think they could be more effective at generalizing their trauma. However, I do not think that it is the responsibility of shrines and shrine builders to make them more effective as stranger-memorials. It is not their job to that. It is ours—the members of the knowing motoring public.

Roadside shrines are working hard to show us something we as the broader motoring public do not yet know how to sense, much less understand. Thus, for too long, the burden of the cultural trauma of living and dying in cars has been carried by individuals as dispersed from each other as they are from the rest of us. But because we all share in the freedom and autonomy afforded by automobility, we should also share the burden of the traumas it produces every day as well.

The complexity of this problem has become most clear to me over time in my many encounters with the site discussed extensively in this chapter, where I have watched the "replaced" Gonzalez girls grow up. Yes, the shrine seems sad to me by itself, but there's something else, another kind of sadness as well. For me, the sadness is connected to the range of affective responses I mentioned earlier. The dolls, especially in close-up, are kitschy, creepy, and poignant all at the same time. I can tell they are trying to do something, but I don't think they are working—at least not the way I think they are "supposed to." But they are doing something. What they are doing is the same thing that every road trauma shrine is doing: interpellating a knowing motoring public and challenging us to do something with them even if—or especially if—we are *continuing to drive.*

For even if we do not know the narrative details, shrines on the side of the road ensure that we remember that people die on the highways doing the exact thing we are doing when we see them: driving, going about the business of living everyday lives, believing in reaching a projected destination—believing, in short, that the future exists. After all, we can see the future up ahead, through the windshield, and we are driving into it, performing freedom and autonomy. But then again, the shrines are there to remind us that other drivers were doing the same thing when they died as well.

In working on this book, I have come to know that the melancholy induced by witnessing the embedded memory of road trauma in the landscape is a powerful force. *Road Scars* represents but hopefully also performs my search for a way to mobilize that force—not only so that we can know what they are doing on the roadside but also so that we can take up the challenge they make to our own participation in the knowing motoring public. Only through paying attention to the places of automotive trauma long enough and intentionally enough can we begin the slow and painful work of figuring out what it means to live in and through the cars that take us not only where we want to go but also where we don't. The thousands of trauma shrines on the roadside are there to remind us that we have a lot of work to do.

NOTES

1 See Fuyuki Kurasawa, "A Message in a Bottle," *Theory, Culture, and Society* 26, no. 1 (2009): 95.

2 Rico Franses, "Monuments and Melancholia," *Journal for the Psychoanalysis of Culture & Society* 6 (2001): 97–104. For alternate analyses of the Vietnam Veterans Memorial and the AIDS Quilt, see Carole Blair and Neil Michel, "The AIDS Memorial Quilt and the Contemporary Culture of Public Commemoration," *Rhetoric & Public Affairs* 10, no. 4 (2007): 595–626; Douglas Crimp, "Mourning and Militancy," *October* 51 (1989): 3–18; Kristin Hass, *Carried to the Wall: American Memory and the Vietnam Veterans Memorial* (Berkeley: University of California Press, 1998); Erin Rand, "Repeated Remembrance: Commemorating the AIDS Quilt and Resuscitating the Mourned Subject," *Rhetoric & Public Affairs* 10, no. 4 (2007): 655–79; and Marita Sturken, *Tangled Memories: The Vietnam War, the AIDS Epidemic, and the Politics of Remembering* (Berkeley: University of California Press, 1997).

3 Franses, "Monuments and Melancholia," 100.

4 Ibid., 104.

5 Ibid., 98.

6 Sigmund Freud, "Mourning and Melancholia," in *The Freud Reader*, ed. P. Gay, trans. J. Strachey (New York: Norton, 1917, 1989): 584–89.

7 Ibid., 586.

8 Ibid., 587.

9 Ibid., 589.

10 Elizabeth Kübler-Ross, *On Death and Dying* (New York: Macmillan, 1969).

11 Dennis Klass, Phyllis Silverman, and Steven Nickman, eds., *Continuing Bonds: New Understandings of Grief* (London: Taylor & Francis, 1996); see also Tammy Clewell, "Mourning beyond Melancholia: Freud's Psychoanalysis of Loss," *Journal of the American Psychoanalytic Association* 52, no. 1 (Spring 2004): 43–67.

12 Franses, "Monuments and Melancholia," 100.

13 Ibid.

14 Ibid., 101.

15 Ibid., 102.

16 Ibid., 101.

17 Michael Warner, *Publics and Counterpublics* (New York: Zone Books, 2002), 72.

18 Kathleen Stewart, "Worlding Refrains," in *The Affect Theory Reader*, ed. Melissa Gregg and Greg Seigworth (Durham: Duke University Press), 344.

19 See "Should Roadside Memorials Be Banned?" Room for Debate (blog), *New York Times*, July 12, 2009, accessed November 20, 2013, http://roomfordebate.blogs.nytimes.com/2009/07/12/should-roadside-memorials-be-banned.

20 John Wylie, "Landscape, Absence, and the Geographies of Love," *Transactions of the Institute of British Geographers* 34 (2009): 285.

21 See Karen Wells, "Melancholic Memorialisation: The Ethical Demands of Grievable Lives," in *Visuality/Materiality: Images, Objects and Practices*, ed. Gillian Rose and Divya P. Tolia-Kelly (Farnham: Ashgate, 2012), 153–69.

22 Ibid., 155.

23 Ibid.

24 Ibid., 160. See also David Eng and David Kazanjian, eds., *Loss: The Politics of Mourning* (Berkeley: University of California Press, 2002).

25 Wells, "Melancholic Memorialisation," 162.

26 Ibid., 163.

27 Ibid., 166.

28 Ibid., 162.

29 David Eng and David Kazanjian, "Introduction: Mourning Remains," in *Loss: The Politics of Mourning*, ed. David Eng and David Kazanjian (Berkeley: University of California Press, 2002), 5. While many commentators noticed a generalized sense of melancholy permeating American culture post 9/11, this book was written just before September 11, and captures the sense of melancholy that was already present in the culture even then.

30 Ibid., 2–3.

31 Ibid., 3.

32 Judith Butler, "After Loss, What Then?" (Afterword), in *Loss: The Politics of Mourning,* ed. David Eng and David Kazanjian (Berkeley: University of California Press, 2002), 470.

33 Ibid., 468.

34 Ibid., 472.

35 Ibid. Emphasis in original.

Bibliography

Ahmed, Sarah. *The Cultural Politics of Emotion*. London: Routledge, 2004.

Alexander, Jeffrey C., "Toward a Theory of Cultural Trauma." In *Cultural Trauma and Collective Identity*, edited by Jeffrey C. Alexander, Ron Eyerman, Bernhard Giesen, Neil Smelser, and Piotr Sztompka, 1–30. Berkeley: University of California Press, 2004.

Alexander, Jeffrey C., Ron Eyerman, Bernhard Giesen, Neil J. Smelser, and Piotr Sztompka, eds. *Cultural Trauma and Collective Identity*. Berkeley: University of California Press, 2004.

Althusser, Louis. *Lenin and Philosophy and Other Essays*. Translated by Ben Brewster. New York: Monthly Review Press, 2001.

Anaya, Rudolfo, Denise Chavez, and Juan Estevan Arellano. *Descansos: An Interrupted Journey*. Albuquerque: El Norte, 1995.

Apel, Dora. *Memory Effects: The Holocaust and the Art of Secondary Witnessing*. New Brunswick: Rutgers University Press, 2002.

Appadurai, Arjun, ed., *The Social Life of Things: Commodities in Cultural Perspective*. Cambridge: Cambridge University Press, 1986.

Árnason, Arnar, Sigurjón Baldur Hafsteinsson, and Tinna Grétarsdóttir, "Acceleration Nation: An Investigation into the Violence of Speed and the Uses of Accidents in Iceland," *Culture, Theory & Critique* 48, no. 2 (2007): 199–217.

Ashcroft, Eric B., "American Atheists, Inc. v. Davenport: Endorsing a Presumption of Unconstitutionality against Potentially Religious Symbols," *Brigham Young University Law Review* 2012, no. 2 (2012): 371–89.

Askins, Kye, and Matej Blazek, "Feeling Our Way: Academia, Emotions and a Politics of Care," *Social & Cultural Geography* 18, no. 8 (2017): 1086–105.

Austin, J. L., *How to Do Things with Words*, 2nd ed. Cambridge: Harvard University Press, 1975.

Azoulay, Ariella, *Death's Showcase: The Power of Image in Contemporary Democracy*. Cambridge: MIT Press, 2001.

Baer, Ulrich, *Spectral Evidence: The Photography of Trauma*. Cambridge: MIT Press, 2005.

Ballard, J. G., *Crash*. New York: Picador, 1973.

Barrera, Alberto, "Mexican-American Roadside Crosses in Starr County." In *Hecho en Tejas: Texas-Mexican Folk Arts and Crafts*, edited by Joe S. Graham, 278–92. Denton: University of North Texas Press, 1997.

Barthes, Roland, *Camera Lucida: Reflections on Photography*. Translated by Richard Howard. New York: Hill & Wang, 1981.

Beal, Timothy, *Roadside Religion: In Search of the Sacred, the Strange, and the Substance of Faith*. Boston, MA: Beacon Press, 2006.

Bean, Hamilton, "'A Complicated and Frustrating Dance': National Security Reform, the Limits of *Parrhesia*, and the Case of the 9/11 Families," *Rhetoric and Public Affairs* 12, no. 3 (2009): 429–60.

Beckmann, Jörg, "Mobility and Safety," *Theory, Culture & Society* 21, no. 4–5 (2004): 81–100.

Bednar, Robert M., "Denying Denial: Trauma, Memory, and Automobility at Roadside Car Crash Shrines." In *Rhetoric, Remembrance, and Visual Form*, edited by Anne T. Demo and Bradford Vivian, 128–45. London: Routledge, 2011.

———, "Killing Memory: Roadside Memorials and the Necropolitics of Affect," *Cultural Politics* 9, no. 3 (2013): 337–56.

———, "Making Space on the Side of the Road: Towards a Cultural Study of Roadside Car Crash Memorials." In *The World Is a Text*, 3rd ed., edited by Jonathan Silverman and Dean Rader, 497–508. Upper Saddle River, NJ: Pearson, 2009.

———, "Materialising Memory: The Public Lives of Roadside Crash Shrines," *Memory Connection* 1, no. 1 (2011): 18–33.

———, "Placing Affect: Remembering Ordinary Trauma at Roadside Crash Shrines." In *Affective Landscapes in Literature, Art, and Everyday Life*, edited by Christine Berberich, Neil Campbell, and Robert Hudson, 49–67. Farnham: Ashgate, 2015.

Belshaw, John, and Diane Purvey, *Private Grief, Public Mourning: The Rise of the Roadside Shrine in British Columbia*. Vancouver: Anvil Press, 2009.

Benjamin, Walter, *Illuminations: Essays and Reflections*. Edited by Hannah Arendt, translated by Harry Zohn. New York: Schocken Books, 1968.

Bennett, Jane, *Vibrant Matter: A Political Ecology of Things*. Durham, NC: Duke University Press, 2010.

Bennett, Jill, "The Aesthetics of Sense-Memory: Theorizing Trauma through the Visual Arts." In *Regimes of Memory*, edited by Susannah Radstone and Katherine Hodgkin, 27–39. London: Routledge, 2003.

Bennett, Jill, *Empathic Vision: Affect, Trauma and Contemporary Art*. Palo Alto, CA: Stanford University Press, 2005.

Berger, John, *Selected Essays*. New York: Vintage, 2001.

Berger, John, and Jean Mohr. *Another Way of Telling*. New York: Vintage, 1982.

Berlant, Lauren, ed., *Compassion: The Culture and Politics of an Emotion*. London: Routledge, 2004.

———, *The Queen of America Goes to Washington City*. Durham, NC: Duke University Press, 1994.

Blair, Carole, Greg Dickinson, and Brian L. Ott, "Introduction: Rhetoric/Memory/ Place." In *Places of Public Memory: The Rhetoric of Museums and Memorials*, edited by Greg Dickinson, Carole Blair, and Brian L. Ott, 1–54. Tuscaloosa: University of Alabama Press, 2010.

Blair, Carole, and Neil Michel, "The AIDS Memorial Quilt and the Contemporary Culture of Public Commemoration," *Rhetoric & Public Affairs* 10, no. 4 (2007): 595–626.

Blum, Ryan H., "Anxious Latitudes: Heterotopias, Subduction Zones, and the Historical-Spatial Configurations within *Dead Man*," *Critical Studies in Media Communication* 27, no. 1 (2010): 55–66.

Böhm, Steffan, Campbell Jones, Chris Land, and Matthew Paterson, eds. *Against Automobility*. Malden, MA: Blackwell, 2006.

Boudreaux, Corrie, "Public Memorialization and the Grievability of Victims in Cuidad Juárez," *Social Research* 83, no. 2 (2016): 391–417.

Bourdieu, Pierre, *Photography: A Middle-Brow Art*. Translated by Shaun Whiteside. Palo Alto, CA: Stanford University Press, 1991.

Bourgois, Phillipe, and Jeff Schonberg, *Righteous Dope Fiend*. Berkeley: University of California Press, 2009.

Breen, Lauren J., and Moira O'Connor, "Acts of Resistance: Breaking the Silence of Grief Following Traffic Crash Fatalities," *Death Studies* 34, no. 1 (2010): 30–53.

Brottman, Mikita, ed., *Car Crash Culture*. New York: Palgrave, 2001.

Brouwer, Daniel C., and Robert Asen, eds., *Public Modalities: Rhetoric, Culture. Media, and the Shape of Public Life*. Tuscaloosa: University of Alabama Press, 2010.

Brummett, Barry, "Electric Literature as Equipment for Living," *Critical Studies in Mass Communication* 2, no. 3 (1985): 247–61.

Buchli, Victor, "Introduction." In *The Material Culture Reader*, edited by Victor Buchli, 1–22. Oxford: Berg, 2002.

Burchell, Graham, "Liberal Government and Techniques of the Self." In *Foucault and Political Reason: Liberalism, Neo-Liberalism, and Rationalities of Government*, edited by Andrew Barry, Thomas Osborne, and Nikolas Rose, 19–36. Chicago, IL: University of Chicago Press, 1996.

Burchianti, Margaret, "Building Bridges of Memory: The Mothers of the Plaza de Mayo and the Cultural Politics of Maternal Memories," *History and Anthropology* 15, no. 2 (2004): 133–50.

Burke, Kenneth, *The Philosophy of Literary Form: Studies in Symbolic Action*, 3rd ed. Berkeley: University of California Press, 1973.

Butler, Judith, "After Loss, What Then?" (Afterword). In *Loss: The Politics of Mourning*, edited by David Eng and David Kazanjian, 467–73. Berkeley: University of California Press, 2002.

———, *Gender Trouble*. New York: Routledge, 1989.

———, *Notes toward a Performative Theory of Assembly*. Cambridge: Harvard University Press, 2015.

Byrd, Rachael M., "Rest in Place: Understanding Traumatic Death along the Roadsides of the Southwestern United States," *Arizona Anthropologist* 26 (2016): 53–75.

Campbell, Elaine, "Public Sphere as Assemblage: The Cultural Politics of Roadside Memorialization," *British Journal of Sociology* 64, no. 3 (2013): 526–47.

Cann, Candi, *Virtual Afterlives: Grieving the Dead in the Twenty-First Century.* Lexington: University of Kentucky Press, 2014.

Carr, Chantel, and Chris Gibson, "Animated Geographies of Making: Embodied Slow Scholarship for Participant Researchers of Maker Cultures and Material Work," *Geography Compass* 11 (2017): 1–10.

Caruth, Cathy, "Recapturing the Past: Introduction." In *Trauma: Explorations in Memory*, edited by Cathy Caruth, 151–57. Baltimore, MD: Johns Hopkins University Press, 1995.

———, "Trauma and Experience: Introduction." In *Trauma: Explorations in Memory*, edited by Cathy Caruth, 3–12. Baltimore, MD: Johns Hopkins University Press, 1995.

———, *Unclaimed Experience: Trauma, Narrative, and History.* Baltimore, MD: Johns Hopkins University Press, 1996.

Charland, Maurice, "Constitutive Rhetoric: The Case of the *Peuple Québécois*," *Quarterly Journal of Speech* 73 (1987): 133–50.

Chopra, Rohit, "The 26/11 Network Archive: Public Memory, History, and the Global in an Age of Terror," *International Journal of Communication* 9 (2015): 1140–62.

Clark, Jennifer, "Challenging Motoring Functionalism: Roadside Memorials, Heritage and History in Australia and New Zealand," *Journal of Transport History* 29, no. 1 (2008): 23–43.

———, ed., *Roadside Memorials: A Multidisciplinary Approach.* Armidale: Emu Press, 2007.

Clark, Jennifer, and Ashley Cheshire, "R.I.P.: A Comparative Study of Roadside Memorials in New South Wales, Australia and Texas, USA," *Omega* 35, no. 2 (2003–4): 229–48.

Clark, Jennifer, and Majella Franzmann, "Authority from Grief: Presence and Place in the Making of Roadside Memorials," *Death Studies* 30, no. 6 (2006): 579–99.

———, " 'A Father, a Son, My Only Daughter': Memorializing Road Trauma," *RoadWise* 13, no. 3 (2002): 4–10.

Clewell, Tammy, "Mourning beyond Melancholia: Freud's Psychoanalysis of Loss," *Journal of the American Psychoanalytic Association* 52, no. 1 (Spring 2004): 43–67.

Clough, Patricia Ticineto, "Introduction." In *The Affective Turn: Theorizing the Social*, edited by Patricia Ticineto Clough, 1–33. Durham, NC: Duke University Press, 2007.

Cohen, Erik, "Roadside Memorials in Northeastern Thailand," *Omega: Journal of Death & Dying* 66, no. 4 (2012): 343–63.

Collins, Catherine Ann, and Alexandra Opie, "When Places Have Agency: Roadside Shrines as Traumascapes," *Continuum: Journal of Media & Cultural Studies* 24, no. 1 (2010): 107–18.

Collins, Charles, and Charles Rhine, "Roadside Memorials," *Omega: The Journal of Death and Dying* 47, no. 3 (2003): 221–44.

Corkill, Claire, and Ray Moore, " 'The Island of Blood': Death and Commemoration at the Isle of Man TT Races," *World Archaeology* 44, no. 2 (2012): 248–62.

Cresswell, Tim, *On the Move: Mobility in the Modern Western World*. London: Routledge, 2006.

———, *Place: A Short Introduction*. Malden, MA: Wiley/Blackwell, 2004.

Crimp, Douglas, "Mourning and Militancy," *October* 51 (1989): 3–18.

Cvetkovich, Ann, *An Archive of Feelings: Trauma, Sexuality, and Lesbian Public Cultures*. Durham, NC: Duke University Press, 2003.

———, "Public Feelings," *South Atlantic Quarterly* 106, no. 3 (2007): 459–68.

Dahl, Richard, "Vehicular Manslaughter: The Global Epidemic of Traffic Deaths," *Environmental Health Perspectives* 112, no. 11 (2004): 628–31.

Damjanov, Katarina, "Lunar Cemetery: Global Heterotopia and the Biopolitics of Death," *Leonardo* 46, no. 2 (2013): 159–62.

de Certeau, Michel, *The Practice of Everyday Life*. Translated by Steven F. Rendall. Berkeley: University of California Press, 1984.

De Groot, Jocelyn, "Maintaining Relational Continuity with the Deceased on Facebook," *Omega* 65, no. 3 (2012): 195–212.

De León, Jason, *The Land of Open Graves: Living and Dying on the Migrant Trail*. Berkeley: University of California Press, 2015.

de Souza e Silva, Adriana, and Jordan Frith, *Mobile Interfaces in Public Spaces: Locational Privacy, Control, and Urban Sociality*. New York: Routledge, 2012.

DeLanda, Manuel, *Assemblage Theory*. Edinburgh: Edinburgh University Press, 2016.

———, *A New Philosophy of Society: Assemblage Theory and Social Complexity*. New York: Continuum, 2006.

Deleuze, Gilles, and Félix Guatarri, *A Thousand Plateaus: Capitalism and Schizophrenia*. Translated by Brian Massumi. Minneapolis: University of Minnesota Press, 1987.

Derrida, Jacques, *Limited, Inc*. Evanston, IL: Northwestern University Press, 1988.

Dickinson, George E., and Heath Hoffman, "Roadside Memorial Policies in the United States," *Mortality* 15, no. 2 (2010): 154–67.

Dickinson, Greg, Carole Blair, and Brian L. Ott, eds., *Places of Public Memory: The Rhetoric of Museums and Memorials*. Tuscaloosa: University of Alabama Press, 2010.

Dobler, Robert Thomas, "Ghost Bikes: Memorialization and Protest on City Streets." In *Grassroots Memorials: The Politics of Memorializing Traumatic Death*, edited by Peter Jan Margry and Cristina Sánchez-Carretero, 169–87. New York: Berghahn Books, 2011.

Donnan, Hastings, "Material Identities: Fixing Ethnicity in the Irish Borderlands," *Identities: Global Studies in Culture and Power* 12 (2005): 69–105.

Donofrio, Theresa Ann, "Ground Zero and Place-Making Authority: The Conservative Metaphors in 9/11 Families' 'Take Back the Memorial' Rhetoric," *Western Journal of Communication* 74, no. 2 (2010): 150–69.

Doss, Erika, "Death, Art, and Memory in the Public Sphere: The Visual and Material Culture of Grief in Contemporary America," *Mortality* 7, no. 1 (2002): 63–82.

———, "Spontaneous Memorials and Contemporary Modes of Mourning in America," *Material Religion* 2, no. 3 (2006): 294–319.

———, *The Emotional Life of Public Memorials: Towards a Theory of Temporary Memorials*. Amsterdam: Amsterdam University Press, 2008.

———, *Memorial Mania: Public Feeling in America*. Chicago, IL: University of Chicago Press, 2010.

Durbin, Jeffrey L., "Expressions of Mass Grief and Mourning: The Material Culture of Makeshift Memorials," *Material Culture* 35 (2003): 22–47.

Dwyer, Owen J., and Derek H. Alderman, eds., *Civil Rights Memorials and the Geography of Memory*. Chicago, IL: Center for American Places, 2008.

Edensor, Tim, *Industrial Ruins: Spaces, Aesthetics, and Materiality*. New York: Berg, 2005.

Edwards, Elizabeth, "Material Beings: Objecthood and Ethnographic Photographs," *Visual Studies* 17 (2002): 67–75.

Edwards, Elizabeth, Chris Gosden, and Ruth Phillips, eds., *Sensible Objects: Colonialism, Museums, and Material Culture*. Oxford: Berg, 2006.

Edwards, Elizabeth, and Janice Hart, eds., *Photographs Objects Histories: On the Materiality of Images*. London: Routledge, 2004.

Elvebakk, Beate, "Vision Zero: Remaking Road Safety," *Mobilities* 2, no. 3 (2007): 425–41.

Eng, David, and David Kazanjian, eds., *Loss: The Politics of Mourning*. Berkeley: University of California Press, 2002.

Eng, David, and David Kazanjian, "Introduction: Mourning Remains." In *Loss: The Politics of Mourning*, edited by David Eng and David Kazanjian, 1–25. Berkeley: University of California Press, 2002.

Erll, Astrid, *Memory in Culture*. Translated by Sara B. Young. New York: Palgrave MacMillan, 2010.

Everett, Holly, "Roadside Crosses and Memorial Complexes in Texas," *Folklore* 111, no. 1 (2000): 91–118.

———, *Roadside Crosses in Contemporary Memorial Culture*. Denton: University of North Texas Press, 2002.

Eyerman, Ron, *Memory, Trauma, and Identity*. New York: Palgrave MacMillan, 2019.

Farrell, Jeff, "Speed Kills," *Critical Criminology* 11 (2002): 185–98.

Featherstone, Mike, "Automobilities: An Introduction," *Theory, Culture & Society* 21, no. 4–5 (2004): 1–24.

Felman, Shoshana, and Dori Laub, eds., *Testimony: Crises of Witnessing in Literature, Psychoanalysis, and History*. New York: Routledge, 1992.

Foote, Kenneth, *Shadowed Ground: America's Landscapes of Violence and Tragedy*, 2nd ed. Austin: University of Texas Press, 2003.

Foucault, Michel, *The Archeology of Knowledge and the Discourse on Language*. Translated by A. M. Sheridan Smith. New York: Pantheon, 1972.

———, "The Ethic of Care of the Self as a Practice of Freedom: An Interview." In *The Final Foucault*, edited by James Bernauer and David Rasmussen, translated by J. D. Gauthier, 1–20. Cambridge: MIT Press, 1988.

———, "Governmentality." In *The Foucault Effect: Studies in Governmentality*, edited by Graham Burchell, Colin Gordon, and Peter Miller, 87–104. Chicago, IL: University of Chicago Press, 1991.

———, "Of Other Spaces," *Diacritics* 16 (Spring 1986): 22–27.

————, *Power/Knowledge: Selected Interviews and Other Writings, 1972–1977.* Edited by Colin Gordon. New York: Random House, 1988.

Franses, Rico, "Monuments and Melancholia," *Journal for the Psychoanalysis of Culture & Society* 6 (2001): 97–104.

Franck, Karen A., and Quentin Stevens, eds., *Loose Space: Possibility and Diversity in Urban Life.* London: Routledge, 2007.

Freud, Sigmund, "Mourning and Melancholia." In *The Freud Reader*, edited by P. Gay and translated by J. Strachey, 584–89. New York: Norton, [1917] 1989.

Gell, Alfred, *Art and Agency: An Anthropological Theory.* Oxford: Oxford University Press, 1998.

Ghost Bikes Website, http://ghostbikes.org. Accessed August 2, 2019.

Gibson, Margaret, "Death and Grief in the Landscape: Private Memorials in Public Space," *Cultural Studies Review* 17, no. 1 (2011): 146–61.

Gibson, Margaret, *Objects of the Dead: Mourning and Memory in Everyday Life.* Victoria: Melbourne University Press, 2008.

Gilroy, Paul, "Driving While Black." In *Car Cultures*, edited by Daniel Miller, 81–104. Oxford, Berg, 2001.

Godel, Margaret, "Images of Stillbirth: Memory, Mourning, and Memorial," *Visual Studies* 22, no. 3 (2007): 253–69.

Gordon, Avery, *Ghostly Matters: Haunting and the Sociological Imagination*, 2nd ed. Minneapolis: University of Minnesota Press, 2008.

Graham, Connor, Michael Arnold, Tamara Kohn, and Martin Gibbs, "Gravesites and Websites: A Comparison of Memorialisation," *Visual Studies* 30, no. 1 (2015): 37–53.

Grider, Sylvia, "Public Grief and the Politics of Memorial: Contesting the Memory of 'The Shooters' at Columbine High School," *Anthropology Today* 23, no. 3 (2007): 3–7.

————, "Roadside Crosses: Vestiges of Colonial Spain in Contemporary New Mexico." In *Descansos: The Sacred Landscape of New Mexico*, edited by Joan E. Alessi, 11–28. Santa Fe, NM: Fresco Fine Art Publications, 2006.

————, "Spontaneous Shrines: A Modern Response to tragedy and Disaster," *New Directions in Folklore* 5 (2001): 1–10.

Griffith, James, and Francisco Manzo Taylor, "Voices from inside a Black Snake, Part II: Sonoran Roadside *Capillas*," *Journal of the Southwest* 48, no. 3 (2006): 233–59.

Grossberg, Lawrence, "Affect's Future: Rediscovering the Virtual in the Actual." In *The Affect Theory Reader*, edited by Melissa Gregg and Gregory Seigworth, pp. 308–38. Durham, NC: Duke University Press, 2010.

Grusin, Richard, ed., *The Nonhuman Turn.* Minneapolis: University of Minnesota Press, 2015.

Guerin, Frances, and Roger Hallas, *The Image and the Witness: Trauma, Memory and Visual Culture.* London: Wallflower Press, 2007.

Gusfield, Joseph R., *The Culture of Public Problems: Drinking-Driving and the Symbolic Order.* Chicago, IL: University of Chicago Press, 1981.

Hass, Kristin, *Carried to the Wall: American Memory and the Vietnam Veterans Memorial.* Berkeley: University of California Press, 1998.

Halbwachs, Maurice, *On Collective Memory*. Translated and edited by Lewis A. Coser. Chicago, IL: University of Chicago Press, 1992.

Hales, Molly, "Animating Relations, Digitally Mediated Intimacies between the Living and the Dead," *Cultural Anthropology* 34, no. 2 (2019): 187–212.

Hallam, Elizabeth, and Jenny Hockey, *Death, Memory, and Material Culture*. Oxford: Berg, 2001.

Haney, C. Allen, Christina Leimer, and Juliann Lowery, "Spontaneous Memorialization: Violent Death and Emerging Mourning Ritual," *Omega: The Journal of Death and Dying* 35, no. 2 (1997): 159–71.

Hannam, Kevin, Mimi Sheller, and John Urry, "Editorial: Mobilities, Immobilities and Moorings," *Mobilities* 1, no. 1 (2006): 1–22.

Hartig, Kate V., and Kevin M. Dunn, "Roadside Memorials: Interpreting New Deathscapes in Newcastle, New South Wales," *Australian Geographical Studies* 36, no. 1 (1998): 5–20.

Haskins, Ekaterina, and Michael Rancourt, "Accidental Tourists: Visiting Ephemeral War Memorials," *Memory Studies* (May 2016): 1–15.

Hawkins, Gay, "History in Things: Sebald and Benjamin on Transience and Detritus," *Amsterdamer Beiträge zur Neueren Germanistik* 72, no. 1 (2009): 161–75.

Henzel, Cynthia, "Cruces in the Roadside Landscape of Northeastern New Mexico," *Journal of Cultural Geography* 11 (1995): 93–106.

Herman, Judith, *Trauma and Recovery: The Aftermath of Violence from Domestic Abuse to Political Terror*, 2nd ed. New York: Basic Books, 1997.

Highmore, Ben, *Ordinary Lives: Studies in the Everyday*. London: Routledge, 2010.

———, "A Sideboard Manifesto: Design Culture in an Artificial World." In *The Design Culture Reader*, edited by Ben Highmore, 1–11. London: Routledge, 2009.

Hirsch, Marianne, *Family Frames: Photography, Narrative, and Postmemory*. Cambridge: Harvard, 1997.

———, ed., *The Familial Gaze*. Hanover: Dartmouth University Press, 1999.

Hoffman, Danny, *Monrovian Modern*. Durham, NC: Duke University Press, 2017.

Holloway, Margaret, Miraslava Hukelova, and Louis Bailey, *Remember Me: Memorialisation in Contemporary Society* (Hull: University of Hull, 2018), https://remembermeproject.wordpress.com. Accessed January 12, 2019.

Hubbard, Phil, and Rob Kitchin, eds., *Key Thinkers on Space and Place*. London: Sage, 2010.

Huyssen, Andreas, *Present Pasts: Urban Palimpsests and the Politics of Memory*. Palo Alto, CA: Stanford University Press, 2003.

Iveson, Kurt, *Publics and the City*. Oxford: Blackwell, 2007.

Jakob, Joey Brooke, "What Remains of Abu Ghraib?: Digital Photography and Cultural Memory," *Visual Studies* 31, no. 1 (2016): 22–33.

Johnson, Peter, "Unraveling Foucault's 'Different Spaces,'" *History of the Human Sciences* 19, no. 4 (2006): 75–90.

Kansteiner, Wulf, "Genealogy of a Category Mistake: A Critical Intellectual History of the Cultural Trauma Metaphor," *Rethinking History* 8, no. 2 (2004): 193–221.

Kansteiner, Wulf, and Harald Weilnböck, "Against the Concept of Cultural Trauma." In *Cultural Memory Studies: An Interdisciplinary and International Handbook*, edited by Astrid Erll and Ansgar Nünning, 229–41. New York: Walter de Gruyter Press, 2008.

Kaplan, E. Ann, *Trauma Culture: The Politics of Terror and Loss in Media and Literature*. New Brunswick: Rutgers University Press, 2005.

Kearl, Michael C., "The Proliferation of Postselves in American Civic and Popular Cultures," *Mortality* 15, no. 1 (2010): 47–63.

Keightley, Emily, and Michael Pickering, "Painful Pasts." In *Research Methods for Memory Studies*, edited by Emily Keightley and Michael Pickering, 151–66. Edinburgh: University of Edinburgh Press, 2013.

Kellaher, Leonie, and Ken Worpole, "Bringing the Dead Back Home: Urban Public Spaces as Sites for New Patterns of Mourning and Memorialization." In *Deathscapes: Spaces for Death, Dying, Mourning, and Remembrance*, edited by Avril Maddrell and James D. Sidaway, 161–80. Lanham: Ashgate, 2010.

Kennerly, Rebecca, "Getting Messy: In the Field and at the Crossroads with Roadside Shrines," *Text/Performance Quarterly* 22 (2002): 229–60.

———, "Locating the Gap between Grace and Terror: Performative Research and Spectral Images of (and on) the Road," *FQS/Forum: Qualitative Research* 9, no. 2 (2008), n.p., http://www.qualitative-research.net/index.php/fqs/article/view/396.

Klaassens, Mirjam, and Maarten J. Bijlsma, "New Places of Remembrance: Individual Web Memorials in the Netherlands," *Death Studies* 38, no. 5 (2014): 283–93.

Klaassens, Mirjam, Peter Groote, and Paulus P. P. Huigen, "Roadside Memorials from a Geographical Perspective," *Mortality* 14, no. 2 (2009): 187–220.

Klaassens, Mirjam, Peter D. Groote, and Frank VanClay, "Expressions of Private Mourning in Public Space: The Evolving Structure of Spontaneous and Permanent Roadside Memorials in the Netherlands," *Death Studies* 37, no. 2 (2013): 145–71.

Klass, Dennis, and Edith Maria Steffen, eds., *Continuing Bonds in Bereavement: New Directions in Research and Practice*. London: Routledge, 2017.

Klass, Dennis, Phyllis R. Silverman, and Steven L. Nickman, eds., *Continuing Bonds: New Understandings of Grief*. London: Taylor & Francis, 1996.

Knappett, Carl, *Thinking through Material Culture: An Interdisciplinary Perspective*. Philadelphia: University of Pennsylvania Press, 2005.

Knight, Kelvin, "Placeless Places: Resolving the Paradox of Foucault's Heterotopia," *Textual Practice* 31, no. 1 (2017): 141–58.

Kopytoff, Igor, "The Cultural Biography of Things: Commoditization as Process." In *The Social Life of Things: Commodities in Cultural Perspective*, edited by Arjun Appadurai, 64–94. Cambridge: Cambridge University Press, 1986.

Kozak, David, "Dying Badly: Violent Death and Religious Change among the Tohono O'odham," *Omega: Journal of Death and Dying* 23 (1991): 2017–216.

Kozak, David, and Camillus Lopez, "The Tohono O'odham Shrine Complex: Memorializing the Location of Violent Death," *New York Folklore* 17, no. 1–2 (1991): 10–20.

Kress, Gunther, and Theo van Leeuwen, *Reading Images: The Grammar of Visual Design*, 2nd ed. London: Routledge, 2006.

Kübler-Ross, Elizabeth, *On Death and Dying*. New York: Macmillan, 1969.

Kurasawa, Fuyuki, "A Message in a Bottle," *Theory, Culture, and Society* 26, no. 1 (2009): 92–111.

LaCapra, Dominick, *Writing History, Writing Trauma*. Baltimore, MD: Johns Hopkins Press, 2001.

Ladd, Brian, *Autophobia: Love and Hate in the Automotive Age*. Chicago, IL: University of Chicago Press, 2008.

Landsberg, Alison, *Prosthetic Memory: The Transformation of American Remembrance in the Age of Mass Culture*. New York: Columbia University Press, 2004.

Langford, Martha, *Suspended Conversations: The Afterlife of Memory in Photographic Albums*. Montreal: McGill-Queen's University Press, 2001.

Layne, Linda L., *Motherhood Lost: The Cultural Construction of Pregnancy Loss in the United States*. New York: Routledge, 2002.

Lee, Rebekah, "Death in Slow Motion: Funerals, Ritual Practice and Road Danger in South Africa," *African Studies* 71, no. 2 (2012): 195–211.

Lefebvre, Henri, *The Production of Space*. Translated by Donald Nicholson-Smith. Oxford: Blackwell, 1991.

Lennon, John, and Malcolm Foley, *Dark Tourism: The Attraction of Death and Disaster*. Andover: Cengage, 2010.

Leys, Ruth, *Trauma: A Genealogy*. Chicago, IL: University of Chicago Press, 2000.

Linenthal, Edward T., *The Unfinished Bombing: Oklahoma City in American Memory*. New York: Oxford University Press, 2003.

Liss, Andrea, *Trespassing through Shadows: Memory, Photography and the Holocaust*. Minneapolis: University of Minnesota Press, 1998.

Luckhurst, Roger, "Traumaculture," *New Formations* 50 (2003): 28–47.

MacConville, Una, "Roadside Memorials: Making Grief Visible," *Bereavement Care* 29, no. 3 (2010): 34–36.

MacConville, Una, and Regina McQuillan, "Remembering the Dead: Roadside Memorials in Ireland," *At the Interface/Probing the Boundaries* 58 (2009): 135–55.

Maddrell, Avril, "Living with the Deceased: Absence, Presence and Absence-Presence," *Cultural Geographies* 20, no. 4 (2012): 501–22.

———, "Online Memorials: The Virtual as the New Vernacular," *Bereavement Care* 31, no. 2 (2012): 46–54.

Maddrell, Avril, and James D. Sidaway, "Introduction: Bringing a Spatial Lens to Death, Dying, Mourning, and Remembrance." In *Deathscapes: Spaces for Death, Dying, Mourning, and Remembrance*, edited by Avril Maddrell and James D. Sidaway, 1–16. Farnham: Ashgate, 2010.

Mahar, Lisa, *American Signs: Form and Meaning on Route 66*. New York: Monacelli Press, 2002.

Margalit, Avishai, *The Ethics of Memory*. Cambridge: Harvard University Press, 2002.

Margry, Peter Jan, and Cristina Sánchez-Carretero, "Memorializing Traumatic Death," *Anthropology Today* 23, no. 3 (2007): 1–2.

———, eds., *Grassroots Memorials: The Politics of Memorializing Traumatic Death*. New York: Berghahn Books, 2011.

Marwick, Alice, and Nicole B. Ellison, "'There Isn't Wifi in Heaven!': Negotiating Visibility on Facebook Memorial Pages," *Journal of Broadcasting & Electronic Media* 56, no. 3 (2012): 378–400.

Massey, Doreen, "Geographies of Responsibility," *Geografiska Annaler Series B: Human Geography* 86, no. 1 (2004): 5–18.

Mazur, Eric, and Kate McCarthy, eds., *God in the Details: American Religion in Popular Culture*. New York: Routledge, 2010.

Mbembe, Achille, "Necropolitics," translated by Libby Meintjes, *Public Culture* 15 (2003): 11–40.

McDannell, Colleen, *Material Christianity: Religion and Popular Culture in America*. New Haven, CT: Yale University Press, 1995.

Meinig, D. W., ed., *The Interpretation of Ordinary Landscapes: Geographical Essays*. Oxford: Oxford University Press, 1979.

Mendel, Maria, "Heterotopias of Homelessness: Citizenship on the Margins," *Studies in Philosophy & Education* 30, no. 2 (2011): 155–68.

Merriman, Peter, "'Mirror, Signal, Manoeuvre': Assembling and Governing the Motorway Driver in Late 1950s Britain." In *Against Automobility*, edited by Steffan Böhm, Campbell Jones, Chris Land, and Matthew Paterson, 75–92. Malden, MA: Blackwell, 2006.

Michael, Mike, "The Invisible Car: The Cultural Purification of Road Rage." In *Car Cultures*, edited by Daniel Miller, 59–80. Oxford: Berg, 2001.

Miller, Daniel, ed., *Car Cultures*. Oxford: Berg/Bloomsbury, 2001.

———, *Stuff*. Malden, MA: Polity Press, 2009.

Miller, Nancy K., and Jason Tougaw, eds., *Extremities: Trauma, Testimony and Community*. Urbana: University of Illinois Press, 2002.

Mills, Katie. *The Road Story and the Rebel: Moving through Film, Fiction, and Television*. Carbondale: Southern Illinois Press, 2006.

Milosevic, Ana, "Remembering the Present: Dealing with the Memories of Terrorism in Europe," *Journal of Terrorism Research* 8, no. 2 (2017): 44–61.

———, "Historicizing the Present: Brussels Attacks and the Heritagization of Spontaneous Memorials," *International Journal for Heritage Studies* 24, no. 1 (2018): 53–65.

Misztal, Barbara. *Theories of Social Remembering*. Maidenhead: Open University Press, 2003.

Mitchell, W. J. T., "Introduction." In *Landscape and Power*, edited by W. J. T Mitchell, 1–4. Chicago, IL: University of Chicago Press, 1994.

———, *Picture Theory*. Chicago, IL: University of Chicago Press, 1994.

———, *What Do Pictures Want? The Lives and Loves of Images*. Chicago, IL: University of Chicago Press, 2005.

Monger, George, "Modern Wayside Shrines," *Folklore* 108 (1997): 113–14.

Morgan, David, *The Embodied Eye: Religious Visual Culture and the Social Life of Feeling*. Berkeley: University of California Press, 2012.

———, *The Lure of Images: A History of Religion and Visual Media in America*. New York: Routledge, 2007.

———, ed., *Religion and Material Culture: The Matter of Belief.* New York: Routledge, 2009.

———, *The Sacred Gaze: Religious Visual Culture in Theory and Practice.* Berkeley: University of California Press, 2005.

Morris, Errol, *Believing Is Seeing: Observations on the Mysteries of Photography.* New York: Penguin, 2011.

Morrison, Toni, *Beloved.* New York: Penguin/Plume, 1988.

Möser, Kurt, "The Dark Side of 'Automobilism,' 1900–1930: Violence, War and the Motor Car," *Journal of Transport History* 24, no. 2 (2003): 238–58.

Müller, Martin, "Assemblages and Actor-Networks: Re-Thinking Socio-Material Power, Politics and Space," *Geography Compass* 9, no. 1 (2015): 27–41.

Nakassis, Constantin, "Citation and Citationality," *Signs and Society* 1, no. 1 (2013): 51–78.

National Highway Traffic Safety Administration, "2016 Fatal Motor Vehicle Crashes: An Overview," https://www.nhtsa.gov/press-releases/usdot-releases-2016-fatal-traffic-crash-data. Accessed April 9, 2018.

———, "Fatality Analysis Reporting System (FARS) Encyclopedia," http://www-fars.nhtsa.gov/Main/index.aspx. Accessed July 5, 2011.

———, "NHTSA Confirms Fatalities Increased in 2012," https://www.nhtsa.gov/press-releases/nhtsa-data-confirms-traffic-fatalities-increased-2012. Accessed April 9, 2018.

———, "Quick Facts 2016," https://www.nhtsa.gov/press-releases/usdot-releases-2016-fatal-traffic-crash-data. Accessed April 9, 2018.

Nesporova, Olga, and Irina Stahl, "Roadside Memorials in the Czech Republic and Romania: Memory versus Religion in Two European Post-Communist Countries," *Mortality* 19, no. 1 (2014): 22–40.

Newbury, Darren, "Making Arguments with Images: Visual Scholarship and Academic Publishing." In *The Sage Handbook of Visual Research Methods*, edited by Eric Margolis and Luc Pauwels, 651–64. London: Sage, 2011.

Newland, Paul, "Look Past the Violence: Automotive Destruction in American Movies," *European Journal of American Culture* 28, no. 1 (2009): 5–20.

Nora, Pierre, "Between Memory and History: Les Lieux de Mémoire," *Representations* 26 (1989): 7–24.

Olick, Jeffrey K., "Collective Memory: The Two Cultures," *Sociological Theory* 17, no. 3 (1999): 333–48.

Ortiz, Carmen, "Pictures That Save, Pictures That Soothe: Photographs at the Grassroots Memorials to the Victims of March 11, 2004 Madrid Bombings," *Visual Anthropology Review* 29, no. 1 (2013): 57–71.

Ott, Brian L., *The Small Screen: How Television Equips Us to Live in the Information Age.* Malden, MA: Wiley-Blackwell, 2007.

Owens, Maida, "Louisiana Roadside Memorials: Negotiating an Emergent Tradition." In *Spontaneous Shrines and the Public Memorialization of Death*, edited by Jack Santino, 119–46. New York: Palgrave, 2006.

Packer, Jeremy, "Disciplining Mobility: Governing and Safety." In *Foucault, Cultural Studies, and Governmentality*, edited by Jack Z. Bratich, Jeremy Packer, and Cameron McCarty, 135–61. Albany, NY: SUNY Press, 2003.

————, *Mobility without Mayhem: Safety, Cars, and Citizenship*. Durham, NC: Duke University Press, 2008.

Paliewicz, Nicholas S., "Bent but Not Broken: Remembering Vulnerability and Resiliency at the National September 11 Memorial Museum," *Southern Communication Journal* 82, no. 1 (2017): 1–14.

Paliewicz, Nicholas S., and Marouf Hasian, Jr., "Mourning Absences, Melancholic Commemoration, and the Contested Public Memories of the National September 11 Memorial and Museum," *Western Journal of Communication* 80, no. 2 (2016): 140–62.

————, "Popular Memory at Ground Zero: A Heterotopology of the National September 11 Memorial and Museum," *Popular Communication* 15, no. 1 (2017): 19–36.

Paton, Nathalie, and Julien Figeac, "Muddled Boundaries of Digital Shrines," *Popular Communication* 13, no. 4 (2015): 251–71.

Perrow, Charles, *Normal Accidents: Living with High-Risk Technologies*. New York: Basic Books, 1984.

Perry, Lieutenant Lee, Phone Interview with the Author, July 21, 2006.

Petersson, Anna, "Swedish *Offercast* and Recent Roadside Memorials," *Folklore* 120, no. 1 (2009): 75–91.

Petersson, Anna, and Carola Wingren, "Designing a Memorial Place: Continuing Care, Passage Landscapes and Future Memories," *Mortality* 16, no. 1 (2011): 54–69.

Pezzullo, Phaedra C., *Toxic Tourism: Rhetorics of Pollution, Travel, and Environmental Justice*. Tuscaloosa: University of Alabama Press, 2007.

Phillips, Kendall R., ed., *Framing Public Memory*. Tuscaloosa: University of Alabama Press, 2004.

Pinney, Christopher, *Camera Indica: The Social Life of Indian Photographs*. Chicago, IL: University of Chicago Press, 1997.

Price, Patricia L., *Dry Land: Landscapes of Belonging and Exclusion*. Minneapolis: University of Minnesota Press, 2004.

Pugliese, Joseph, "Crisis Heterotopias and Border Zones of the Dead," *Continuum: Journal of Media & Cultural Studies* 23, no. 5 (2009): 663–79.

Radford, Gary P., Marie L. Radford, and Jessica Lingel, "The Library as Heterotopia: Michel Foucault and the Experience of Library Space," *Journal of Documentation* 71, no.4 (2015): 733–51.

Radstone, Susannah, and Bill Schwarz, "Mapping Memory." In *Memory: Histories, Theories, Debates*, edited by Susannah Radstone and Bill Schwarz, 1–14. New York: Fordham University Press, 2010.

Rajan, Sudhir Chella, "Automobility, Liberalism, and the Ethics of Driving," *Environmental Ethics* 29, no. 1 (2007): 77–90.

Rand, Erin, "Repeated Remembrance: Commemorating the AIDS Quilt and Resuscitating the Mourned Subject," *Rhetoric & Public Affairs* 10, no. 4 (2007): 655–79.

Redman, Peter, "Affect Revisited: Transference-Countertransference and the Unconscious Dimensions of Affective, Felt, and Emotional Experience," *Subjectivity* 26 (2009): 61–62.

Reid, Jon K., and Cynthia L. Reid, "A Cross Marks the Spot: A Study of Roadside Death memorials in Texas and Oklahoma," *Death Studies* 25 (2001): 341–56.

Reinhardt, Mark, Holly Edwards, and Erina Duganne, eds. *Beautiful Suffering: Photography and the Traffic in Pain.* Chicago, IL: University of Chicago Press, 2007.

Rickert, Thomas, *Ambient Rhetoric: The Attunement of Rhetorical Being.* Pittsburgh: University of Pittsburgh Press, 2013.

Rogers, Everett M., *Diffusion of Innovations.* New York: Glencoe/Free Press, 1962.

Rose, Gillian, *Doing Family Photography: The Domestic, the Public, and the Politics of Sentiment.* London: Ashgate, 2010.

———, "'Everyone's Cuddled Up and It Just Looks Really Nice': An Emotional Geography of Some Mums and Their Family Photos," *Social & Cultural Geography* 5, no. 4 (2004): 549–64.

———, "Family Photographs and Domestic Spacings: A Case Study," *Transactions of the Institute of British Geographers* 28 (2003): 5–18.

———. *Visual Methodologies: An Introduction to Researching with Visual Materials,* 4th ed. London: Sage, 2016.

———, "Who Cares for the Dead and How?: British Newspaper Reporting and the Bombings of London, July 2005," *Geoforum* 40 (2009): 46–54.

Rose, Gillian, and Divya Tolia-Kelly, eds. *Visuality/Materiality: Images, Objects and Practices.* Farnham: Ashgate, 2012.

———, "Visuality/Materiality: Introducing a Manifesto for Practice." In *Visuality/ Materiality: Images, Objects and Practices,* edited by Gillian Rose and Divya Tolia-Kelly, 1–11. Farnham: Ashgate, 2012.

Saindon, Brent Allen, "A Doubled Heterotopia: Shifting Spatial and Visual Symbolism in the Jewish Museum Berlin's Development," *Quarterly Journal of Speech* 98, no. 1 (2012): 4–48.

Saltzman, Lisa, and Eric Rosenberg, eds., *Trauma and Visuality in Modernity.* Hanover: Dartmouth College Press, 2006.

Sandage, Scott A., "A Marble House Divided: The Lincoln Memorial, the Civil Rights Movement, and the Politics of Memory, 1939–1963," *Journal of American History* 80 (1993): 135–67.

Santino, Jack, "Performative Commemoratives: Spontaneous Shrines and the Public Memorialization of Death." In *Spontaneous Shrines and the Public Memorialization of Death,* edited by Jack Santino, 5–16. New York: Palgrave, 2006.

———, "Performative Commemoratives: Spontaneous Shrines, Emergent Ritual, and the Field of Folklore," *Journal of American Folklore* 117 (2003): 363–72.

———, ed., *Spontaneous Shrines and the Public Memorialization of Death.* New York: Palgrave, 2006.

Savage, Kirk. *Monument Wars: Washington, D.C., the National Mall, and the Transformation of the Memorial Landscape.* Berkeley: University of California Press, 2009.

Schivelbusch, Wolfgang. *The Railway Journey: The Industrialization of Time and Space in the 19th Century.* Berkeley: University of California Press, 1986.

Scott, Kimberly, "DATS Trucking Sets 15th Memorial Cross for UHP Trooper on Property Where Atheist Ban Doesn't Apply," *St. George News,* December 30,

2016, https://highwaypatrol.utah.gov/fallen-trooper-memorial/. Accessed February 9, 2017.

Sedgwick, Eve Kosofsky, *Touching Feeling: Affect, Pedagogy, Performativity*. Durham, NC: Duke University Press, 2003.

Seiler, Cotton, *Republic of Drivers: A Cultural History of Automobility in America*. Chicago, IL: University of Chicago Press, 2008.

Seltzer, Mark, *Serial Killers: Death and Life in America's Wound Culture*. New York: Routledge, 1998.

———, "Wound Culture: Trauma in the Pathological Public Sphere," *October* 80 (Spring 1997): 3–26.

Sharpley, Richard, and Philip R. Stone, *The Darker Side of Travel: The Theory and Practice of Dark Tourism*. Bristol: Channel View Press, 2009.

Sheller, Mimi, and John Urry, "The City and the Car," *International Journal of Urban and Regional Research* 24 (2000): 737–57.

———, eds., *Mobile Technologies of the City*. London: Routledge, 2006.

"Should Roadside Memorials Be Banned?" Room for Debate (blog), *New York Times*, July 12, 2009, roomfordebate.blogs.nytimes.com/2009/07/12/should-road side-memorials-be-banned. Accessed November 20, 2013.

Sloop, John M., and Joshua Gunn, "Status Control: An Admonition Concerning the Publicized Privacy of Social Networking." *Communication Review* 13 (2010): 289–308.

Smith, Robert James, "Roadside Memorials: Some Australian Examples," *Folklore* 110, no. 1 (1999): 103–5.

Sontag, Susan, *Regarding the Pain of Others*. New York: Picador, 2003.

Soto, Gabrielle, "Spontaneous Materiality: An Informal Survey of the January 2, 2011 Shrines at University Medical Center," *Arizona Anthropologist* 26 (2016): 76–98.

Stevens, Quentin, and Karen A. Franck, *Memorials as Spaces of Engagement: Design, Uses and Meaning*. London: Routledge, 2016.

Stewart, Kathleen, *Ordinary Affects*. Durham, NC: Duke University Press, 2007.

———, "Worlding Refrains." In *The Affect Theory Reader*, edited by Melissa Gregg and Gregory J. Seigworth, 339–53. Durham, NC: Duke University Press, 2010.

Sturken, Marita, *Tangled Memories: The Vietnam War, The AIDS Epidemic, and the Politics of Remembering*. Berkeley: University of California Press, 1997.

———, *Tourists of History: Memory, Kitsch, and Consumerism from Oklahoma City to Ground Zero*. Durham, NC: Duke University Press, 2007.

Sweeney, Kate, *American Afterlife: Encounters in the Customs of Mourning*. Athens: University of Georgia Press, 2014.

Tay, Richard, "Drivers' Perceptions and Reactions to Roadside Memorials," *Accident Analysis & Prevention* 41, no. 4 (2009): 663–69.

Tay, Richard, Anthony Churchill, and Alexandre C de Barros, "Effects of Roadside Memorials on Traffic Flow," *Accident Analysis and Prevention* 43 (2011): 483–86.

Taylor, Diana, *The Archive and the Repertoire: Performing Culture in the Americas*. Durham, NC: Duke University Press, 2003.

Terranova, Charissa N., *Automotive Prosthetic: Technological Mediation and the Car in Conceptual Art*. Austin: University of Texas Press, 2014.

Tolia-Kelly, Divya, "Locating Processes of Identification: Studying the Precipitates of Re-Memory in the British Asian Home," *Transactions of the Institute of British Geographers* 29 (2004): 314–29.

Torres Smith, Anita, "Descansos: Markers to Heaven," *Journal of Big Bend Studies* 12 (2000): 259–70.

Traverso, Antonio, and Mick Broderick, "Interrogating Trauma: Towards a Critical Trauma Studies," *Continuum: Journal of Media & Cultural Studies* 24, no. 1 (2010): 3–15.

Trujillo, Michael, *Land of Disenchantment: Latina/o Identities and Transformations in Northern New Mexico*. Albuquerque: University of New Mexico Press, 2009.

Tuan, Yi-Fu, *Space and Place: The Perspective of Experience*. Minneapolis: University of Minnesota Press, 1977.

Ulmer, Gregory, "Traffic of the Spheres: Prototype for a Memorial." In *Car Crash Culture*, edited by Mikita Brottman, 327–44. New York: Palgrave, 2001.

Urry, John, *Mobilities*. London: Polity Press, 2007.

———, *Sociology beyond Societies: Mobilities for the Twenty-First Century*. London: Routledge, 2000.

———, "The 'System' of Automobility," *Theory, Culture & Society* 21 (2004): 25–39.

Van Der Kolk, Bessel A., and Onno Van Der Hart, "The Intrusive Past: The Flexibility of Memory and the Engraving of Trauma." In *Trauma: Explorations in Memory*, edited by Cathy Caruth, 158–82. Baltimore, MD: Johns Hopkins University Press, 1995.

Vansina, Jan, *Oral Tradition as History*. Madison: University of Wisconsin Press, 1985.

Veil, Shari R., Timothy L. Sellnow, and Megan Heald, "Memorializing Crisis: The Oklahoma City National Memorial as Renewal Discourse," *Journal of Applied Communication Research* 39, no. 2 (2011): 164–83.

Venturi, Robert, Denise Scott Brown, and Seven Izenour, *Learning from Las Vegas: The Forgotten Symbolism of Architectural Form*. Cambridge: MIT Press, 1977.

Virilio, Paul, and Sylvère Lotringer, *The Accident of Art*. Translated by Michael Taormina. New York: Semiotexte, 2005.

Walter, Tony, Rachid Hourizi, Wendy Moncour, and Stacey Pitsillides, "Does the Internet Change How We Die and Mourn?: Overview and Analysis," *Omega* 64, no. 4 (2011): 275–302.

Warner, Michael, *Publics and Counterpublics*. New York: Zone Books, 2002.

———, "Publics and Counterpublics (Abbreviated Version)," *Quarterly Journal of Speech* 88, no. 4 (2002): 413–25.

Wells, Karen, "Melancholic Memorialisation: The Ethical Demands of Grievable Lives." In *Visuality/Materiality: Images, Objects and Practices*, edited by Gillian Rose and Divya P. Tolia-Kelly, 153–69 Farnham: Ashgate, 2012.

Williams, Paul, *Memorial Museums: The Global Rush to Commemorate Atrocities*. Oxford: Berg, 2007.

Williams, Raymond, *Marxism and Literature*. Oxford: Oxford University Press, 1977.

———, *Towards 2000*. London: Hogarth Press, 1983.

Wollen, Peter, and Joe Kerr, eds., *Autopia: Cars and Culture*. London: Reaction, 2002.

Wouters, Cas, "The Quest for New Rituals in Dying and Mourning: Changes in the We-I Balance," *Body & Society* 8, no. 1 (2002): 1–27.

Wright, Elizabethada, "Rhetorical Spaces in Memorial Spaces: The Cemetery as Rhetorical Memory Place/Space," *Rhetoric Society Quarterly* 35, no. 4 (2005): 51–81.

Wylie, John, "Landscape, Absence, and the Geographies of Love," *Transactions of the Institute of British Geographers* 34 (2009): 275–89.

Young, Craig, and Duncan Light, "Interrogating Spaces of and for the Dead as 'Alternative Space': Cemeteries, Corpses and Sites of Dark Tourism," *International Review of Social Research* 6, no. 2 (2016): 61–72.

Young, James, *The Texture Of Memory: Holocaust Memorials and Meaning*. New Haven, CT: Yale University Press, 1993.

Yudinka, Anna, and Anna Sokolova, "Roadside Memorials in Contemporary Russia: Folk Origins and Global Trends," *Religion & Society in Central & Eastern Europe* 7, no. 1 (2014): 35–51.

Zelizer, Barbie, *About to Die: How News Images Move the Public*. Oxford: Oxford University Press, 2010.

Index

About the Author/Photographer

Robert Matej Bednar is associate professor of communication studies at Southwestern University in Georgetown, Texas, United States, where he teaches courses in media studies, visual/material communication, critical/ cultural studies, and automobility. His work as an interdisciplinary analyst, photographer, and theorist focuses on the performative dimensions of everyday communicative behavior, particularly the ways that people perform identities visually, materially, and spatially as they negotiate public spaces and on the ways that objects, pictures, and spaces communicate beyond representation. He has published a number of scholarly and popular articles and book chapters on road trauma shrines and U.S. National Park snapshot photography practices. *Road Scars* is his first book.

Often, after Professor Bednar has presented his work in public, audience members not only critically engage his arguments and ideas but also share stories of their own road trauma or of seeing trauma shrines on the roadside. Many have e-mailed him photographs of shrines they encounter as drivers. Some even have e-mailed him photographs of shrines they have built or contributed to for people they have lost. Professor Bednar is always grateful to receive this kind of engagement, so please contact him at bednarb@southwestern.edu.

www.ingramcontent.com/pod-product-compliance
Lightning Source LLC
Chambersburg PA
CBHW071848270326
41929CB00013B/2147